JANICE VANCLEAVE'S

A+ Science Fair Workbook and Project Journal

Grades 7–12

WILEY

John Wiley & Sons, Inc.

Published by John Wiley & Sons, Inc., Hoboken, New Jersey
Published simultaneously in Canada

Design and production by Navta Associates, Inc.

Portions of this book were previously published in *Janice VanCleave's A+ Projects in Astronomy,* © 2002 Janice VanCleave; *Janice VanCleave's A+ Projects in Biology,* © 1993 Janice VanCleave; *Janice VanCleave's A+ Projects in Chemistry,* © 1993 Janice VanCleave; *Janice VanCleave's A+ Projects in Earth Science,* © 1999 Janice VanCleave; *Janice VanCleave's A+ Projects in Physics,* © 2003 Janice VanCleave; and *Janice VanCleave's A+ Science Fair Projects,* © 2003 Janice VanCleave; all published by John Wiley & Sons, Inc.

The publisher and the author have made every reasonable effort to insure that the experiments and activities in the book are safe when conducted as instructed but assume no responsibility for any damage caused or sustained while performing the experiments or activities in this book. Parents, guardians, and/or teachers should supervise young readers who undertake the experiments and activities in this book.

For general information about our other products and services, please contact our Customer Care Department within the United States at (800) 762-2974, outside the United States at (317) 572-3993 or fax (317) 572-4002.

Wiley also publishes its books in a variety of electronic formats. Some content that appears in print may not be available in electronic books. For more information about Wiley products, visit our web site at www.wiley.com.

ISBN 0-471-46719-7

Contents

Physics

How to Use This Book

cience is a system of knowledge as well as the pursuit of knowledge about the nature of things in the universe. It is the result of observing, questioning, and experimenting to test one's ideas. A **science fair project** is an investigation that is designed to find the answer to one specific science question. Because science fair projects are done for a contest called a **science fair** (an organized contest in which science fair projects are compared and judged on the basis of predetermined criteria), you must follow certain rules in preparing your project. One of the main requirements is that your project must follow the scientific method. You'll find information on this as well as other basic rules generally required of all science fair projects in this book, but you should also consult your teacher and fair director about the specific rules for your local fair before you start planning your project.

Developing a science fair project is like being a detective. It requires that you plan well, carefully collect facts, and then share your findings. Preparing and presenting a science fair project can be a fun and rewarding experience. But trying to develop the project and/or assemble the display at the last minute usually results in frustration.

Start your project with curiosity and a desire to learn something new. Then proceed with a purpose and a determination to solve the problem selected. Even if your project doesn't turn out exactly as you planned, it is likely that your scientific quest will end with some interesting discoveries.

This book will give you guidance in developing your project from start to finish, including information on the basic scientific tools and techniques needed to develop and present a project. Before you start your project, read all of Part I. It explains nine important points you need to know for science fair success:

1. *The scientific method.* A systematic way of gathering and pursuing scientific knowledge is what the **scientific method** is all about. Chapter 1 describes the steps of the scientific method and how they can be used in developing a science fair project.

2. *Topic research.* Selecting a topic is the first hurdle to jump in getting started on your project. The research suggestions in Chapter 2 will help you choose a topic. **Research** is the process of collecting information and data. **Data,** as used in this book, are observations and/or measured facts obtained experimentally. **Topic research,** as defined in this book, is research done with the objective of selecting a **project topic** (subject of a science fair project).

3. *Categories.* You should identify the category in which your project falls at the beginning of your research. Judges base their evaluation of the content of your project on the category in which you enter it. For example, an A+ earth science project incorrectly entered in the chemistry category most likely will receive a lower rating. Chapter 3 provides a list of some common categories that are used in science fairs.

4. *Project research.* Once a topic has been selected, research begins with the objective of identifying a **project problem,** which is the scientific question to be solved. **Project research** is an in-depth study of the project topic with the objective of expressing a project problem, proposing a **hypothesis** (a prediction to the solution to a problem, based on knowledge and research), and designing a **project experiment** (an experiment designed to test the hypothesis) to test the hypothesis. This research should involve more than just reading printed materials. It should include interviews of people knowledgeable about the topic as well as exploratory experiments. **Exploratory experiments** as defined in this book, are experiments in which the data are used as part of the project research. Chapter 4 provides suggestions and directions for gathering project research.

5. *Project overview.* Chapter 5 provides a brief overview of a specific science fair project from start to finish. This chapter guides you step-by-step through the collection of research and its use in identifying a project problem, proposing a hypothesis, and designing a project experiment. The objective of the chapter is to provide a model of how you can organize and plan your science fair project.

6. *Experimental data.* Data are observations and/or measured facts obtained experimentally. Chapter 6 provides ways to organize and display data,

including examples of different types of graphs. This chapter includes information about the **project journal** (written progress record of your science fair project).

7. *Project summaries.* Chapter 7 includes information on how to prepare an **abstract** (brief overview of a scientific project) and a **project report** (written report of an entire project from start to finish).

8. *The display.* Your display is your way of communicating to others what you did and what you learned. It's important that you use the space you're given wisely to represent your project in the best way possible. Chapter 8 provides ideas for designing a project display that will not only catch the eye of observers but hold their attention.

9. *Presentation and evaluation.* Although your presentation and evaluation come at the end of the process, you should be thinking about them throughout the development of the project. It is important to consider how you will be evaluated so that your project entry meets the necessary criteria. Chapter 9 helps prepare you to be judged and gives hints of what to expect at the fair. Your teacher can provide more specific information.

Part II provides A+ project ideas for 55 science fair topics. An A+ idea is not a guarantee that you will receive an A+ grade on your project. Your grade will depend on you. You must do your part by designing your project, investigating, and collecting and organizing the data to the best of your ability. The point is not to copy the projects in this book but to use them as exploratory experiments and as models in developing your own project. The objective of a science fair project is for you to make your own discoveries. Whether or not you win a ribbon at the fair, all science fair participants who do their best earn the prize of being first-place scientists. This is your opportunity to be a winner! Let's get started!

I

A GUIDE TO SCIENCE FAIR PROJECTS

Chapter 1

The Scientific Method

science fair project requires the use of a set of steps called the scientific method to discover the answer to a scientific problem. This chapter uses examples to illustrate and explain the basic steps of the scientific method. Chapters 2 through 4 give more details about these steps, and Chapter 5 uses the scientific method in an overview of a sample project.

The steps needed in developing a science fair project are, in this order: doing topic and project research, identifying the project problem, stating a hypothesis, conducting project experimentation, and reaching a conclusion. It is important to note that scientists don't always follow these exact steps or do them in this order. However, scientists do follow some type of step-by-step method, using some or all of the steps.

Research

Research is the process of collecting information about a subject. It is also the facts collected, which include information from personal experiences; information from knowledgeable sources, including printed works and interviews with professionals; and observational information from exploratory experiments. Your preliminary research, called topic research, is used to select a project topic. For example, you observe that grass growing around a brick is green, but when you move the brick, the grass under it is yellow. This makes you wonder why the grass changed color. Your topic will be about what makes grass change color. You may choose botany as your overall topic, since this is the study of plants.

Once the topic is selected, you begin what is called project research. This research will help you understand the topic, express a problem, propose a hypothesis, and design a project experiment. From your research into the loss of coloring in grass, you discover that grass is green because of the presence of a green **pigment** (coloring substance) called **chlorophyll.** You also discover that light affects the production of chlorophyll, and the loss of the green color in plants due to the loss of chlorophyll is called **chlorosis.**

Another type of research is an exploratory experiment. Examples of such experiments are given in Part II of this book. Exploratory experiments are not to be confused with the project experiment. Instead they provide information about your topic that helps you to identify the project problem and may give you ideas for your science fair project experiment. An example of an exploratory experiment for the grass color topic would be to place an object on top of a section of grass and observe the color of the grass under the object over a period of time. The results of other project research and this exploratory experiment give you the needed information for the next step—identifying a project problem.

Research Hints

Do use many references from printed sources with current dates like books, journals, magazines, and newspapers, as well as electronic sources such as computer software and on-line sources.

Do gather information from professionals—instructors, librarians, and scientists, such as researchers, physicians, nurses, and veterinarians.

Do perform exploratory experiments, such as those in the 55 science project ideas in Part II.

Project Problem

A project problem is the scientific question for a science project that will be solved. This question is the foundation of your whole project, so it's very important to be specific about what you will be investigating. It must identify the independent and dependent variables. **Variables** are the changing **factors** (conditions contributing to the results of an experiment) in an experiment. The **independent variable** is the factor in an experiment that you purposely manipulate (change). The **dependent variable** is the factor being observed in an experiment that changes in response to the independent variable. Always make sure there is only one independent variable, so you'll know what causes any change you observe in the one dependent variable. There is no right or wrong way of expressing the project problem, but an "open-ended" question, which is answered with a statement, not just a yes or a no, can increase the complexity of the project experiment required to answer it. This makes the project more acceptable for entry in the senior division of a science fair. For example, an open-ended problem for your grass color project might be "What effect does the absence of sunlight have on chlorosis of lawn grass?"

Problem Hints

Do limit your problem. Note that the previous question has one independent variable—light—and one dependent variable—chlorosis. To find the answer to a question such as "How does the absence of light affect the chlorosis of grass?" would require that you test different kinds of light and an extensive variety of grasses. Identifying the type of light and the kind of grass further limits the topic, making it easier to find the answer to the problem. You might even add the name of a specific grass used, such as St. Augustine grass.

Do choose a problem that can be solved experimentally with measurable results. For example, the question "What is the cause of chlorosis?" is a question that can be answered by finding the word *chlorosis* in the dictionary. But "What effect does the absence of sunlight have on the chlorosis of lawn grass?" is a question that can be answered by experimentation. The degree of chlorosis can be measured.

Hypothesis

A hypothesis is a prediction of the solution to a problem, based on knowledge and research. All of your project research is done with the goal of expressing a problem, proposing an answer to it (the hypothesis), and designing a project experiment to test the hypothesis. The hypothesis should make a claim about how the independent and dependent variables relate.

There are two types of science fair hypotheses: a research hypothesis and a null hypothesis. A **research hypothesis** is the expected result of an experiment. It is a specific and testable prediction, and must contain the independent and dependent variables. It can be a declarative statement or an "If . . . then . . ." type of

statement. Here are two examples of a research hypothesis for the sample project question:

"Chlorosis of lawn grass increases as the time without sunlight increases. This is based on the fact that green grass turns yellow if covered by an opaque object for a time."

OR

"If lawn grass does not receive sunlight, then the degree of chlorosis increases with time. This is based on the fact that green grass turns yellow if covered by an opaque object for a time."

A **null hypothesis** is a statement of no predicted change as to the outcome of an experiment. It is often the opposite of what you believe. Your goal is to experimentally reject or fail to reject the null hypothesis. Here is an example of a null hypothesis for the sample project question:

"The absence of sunlight does not affect chlorosis of lawn grass."

Hypothesis Hints

Do state facts from research, including past experiences or observations on which you based your research hypothesis.

Do write down your hypothesis before beginning the project experimentation.

Don't change your hypothesis even if experimentation does not support it. If time permits, repeat or redesign the experiment to confirm your results.

Project Experimentation

A project experiment is an experiment designed to test a hypothesis. For example, you want to test the effect of the absence of sunlight on chlorosis. The independent variable for the experiment is absence of light and the dependent variable is the color of the grass (which indicates the amount of chlorosis). *Note:* the dependent variable needs to be measureable. The factors that are not changing are called **constant factors** or **constants.** The constants should include the type of grass, the time of year, the time of day, atmospheric conditions, and so on. It's important to think of all the possible constants when setting up your experiment to be sure they won't change. A possible experiment to compare the effect of the absence of sunlight over a period of time on the chlorosis of lawn grass would be to cover 10 areas of grass with an **opaque** (light cannot pass through) material, such as opaque plastic cups. Secure the cups by partially pushing them into the ground. Remove one cup every 24 hours and observe the grass that was covered by the cup as well as the uncovered grass around the cup. Take a photograph of the grass that the cup covered. Take a photograph of the grass not covered by the cup for comparison. Repeat this procedure until all 10 cups have been removed. The photographs can be compared to determine the degree of chlorosis with a measured amount of absence of sunlight.

A **control** is a standard for comparison. In a control, all constant factors are identical to the experimental setup (cups placed over grass). Only the independent variable is changed from that of the experimental setup. In the experiment, the independent variable is the absence of light, so a control with full light is needed to determine if in fact it is the absence of light that causes chlorosis. Ten **transparent** (light passes through) cups could be used to allow light to reach the grass. Like the opaque cups in the experiment, the transparent cups would cover the grass, and every 24 hours one of the control cups would be lifted and the grass photographed.

Note that in some experiments, a control is not a separate experimental setup. Instead it can be the independent variable selected as the basis for comparison. For example, if the problem is "How does the angle of the Sun's rays at noon affect seasonal temperature?" the experiment would be to determine air temperature and Sun angle at noon during different seasons. Using the data, the control can arbitrarily be determined to be the Sun at its midpoint angle. Then the temperature for Sun angles greater than or less than the control angle can be compared to the temperature of the control angle.

Project experiments include **observations** (information about events or things). The two types of observations are quantitative and qualitative. A **quantitative observation** is a description of the amount of something. Numbers are used in a quantitative description. Instruments such as a scale, a ruler, or a timer are used to make quantitative descriptions. This means that the amount of the property being observed, such as mass, height, or time is described. Metric measurements are generally the preferred units of measurement for science fair projects, for example, the quantitative observation of length in meters, mass in grams, volume in milliliters, and temperature in Celsius.

A **qualitative observation** is a description of the physical properties of something, including how it looks, sounds, feels, smells, and/or tastes. Adjectives are used in a qualitative description. The qualitative description of grass might relate to its color and include words such as green, yellow, and pale.

Experimentation Hints

When designing your project experiment, remember:

Do include a way of measuring the results. For example, to measure the amount of chlorosis, you can create a color scale from 1 to 10 that measures the degree of color change the grass undergoes. Be sure to specify what 1 to 10 represent, as in: where 1 equals dark green and 10 equals white.

Do use only one independent variable during the experiment.

Do collect more than one set of data to verify your results. Note that for the sample experiment, four or more setups could be started at the same time. This allows the constant factors to be the same for each experiment.

Do have a control.

Do repeat the control along with the rest of the experiment.

Do carefully record and organize the data from your experiment. (See Chapter 6, "Experimental Data," for information on organizing data from an experiment.)

Do have adult supervision for safety assurances.

Project Conclusion

The **project conclusion** is a summary of the results of the project experiment and a statement of how the results relate to the hypothesis. If applicable, you should include reasons for experimental results that are contrary to the hypothesis. The conclusion can end with ideas for further testing. *If your results do not support your hypothesis, you still have accomplished a successful science fair project. Remember that an experiment is done to test the hypothesis. Your experimental results will either support or not support your hypothesis. So:*

Don't change your hypothesis.

Don't leave out experimental results that do not support your hypothesis.

Do give possible reasons for the difference between your hypothesis and the experimental results.

Do make note of ways that you might experiment further to confirm the results of your original experiment. These ideas can be recorded as ideas for future experimenting.

If your results do not support your hypothesis, you might say:

"As stated in the hypothesis, if lawn grass does not receive sunlight, then the degree of chlorosis increases with time. The experimental results did not clearly support the idea that without sunlight, chlorosis increases with time. While there was some chlorosis under the opaque cups, there was no clear-cut evidence of an increase in chlorosis with time. These results may have been due to the cups that I considered to be opaque. In an effort to control constants, the same transparent cups were used, but for the opaque cups, a paper lining was used. For further testing, I would use a different type of paper lining or a cup made of opaque material."

If your results do support your hypothesis you might say:

"As stated in the hypothesis, if lawn grass does not receive sunlight, then the degree of chlorosis increases with time. The data obtained from experimentation support the idea that the absence of sunlight causes an increase in chlorosis. The grass covered for the longest period of time by the opaque cups had the greatest amount of chlorosis and the grass covered by the transparent cup in the control had no chlorosis. For further testing, I would select light as the independent variable and test the effect of different types of light, such as sunlight, fluorescent light, and incandescent light, on chlorosis."

Chapter 2
Topic Research

ow that you have the basic tools for doing a science fair project—the systematic steps of the scientific method—you are ready to learn about the specifics of developing a science fair project.

Project Journal

Your project journal is a written record of the progress of your science fair project. You can use the journal pages at the end of this book, or you can buy a loose-leaf binder with tabbed dividers and some pocket pages. Every journal entry should be as neat as possible and dated. A neat, orderly journal provides a complete and accurate record of your project from start to finish for judges, and it can be used to write your project report. It is also proof of the time you spent searching out the answers to the problem you undertook to solve.

You will want a special section in the journal for handouts provided by your teacher, including the science fair's rules and regulations as well as the timetable that lists when different parts of the project are due. The journal should contain notes on your topic and project research, not only your original ideas but also ideas you get from printed sources or from people. It should also include descriptions of your exploratory and project experiments as well as diagrams, graphs, and written observations of all your results. If it is allowed at your science fair, display the journal with your completed project.

Selecting a Topic

Obviously you want to get an A+ on your project, win awards at the science fair, and learn many new things about science. Some or all of these goals are possible, but you will have to spend a lot of time working on your project, so choose a topic that interests you. It is best to pick a topic and stick with it, but if you find after some work that your topic is not as interesting as you originally thought, stop and select another one. Since it takes time to develop a good project, however,

it is unwise to repeatedly jump from one topic to another. You may, in fact, decide to stick with your original idea even if it is not as exciting as you had expected. You might just uncover some very interesting facts that you didn't know.

Remember that the main objective of a science project is to learn more about science. Your project doesn't have to be highly complex to be successful. Excellent projects can be developed that answer very basic and fundamental questions about events or situations encountered on a daily basis that relate to science. There are many easy ways of selecting a topic. The following are just a few of them.

Ask Questions about the World around You

You can turn everyday experiences into a project topic by asking **inquiry questions** (introductory problems for investigation about a topic). For example, you may have seen a potted plant sitting near a window looking healthy in the morning, but the same plant looking wilted in the afternoon. If you express this as an inquiry question—"What causes the stems of a plant to be stiff?"—you have a good question about plants. But could this be a project topic? Think about it! What is it in the plant that makes it stiff? Did the Sun have any effect on the plant's stiffness? By continuing to ask inquiry questions, you zero in on the topic of water movement through plants via **xylem tubes** (plant structures that transport water from the roots to other plant parts) as well as the cause of **turgor pressure** (internal pressure of plant cells due to the presence of water) in plants.

Keep your eyes and ears open, and start asking yourself more inquiry questions, such as "Do certain types of dyes fade faster than others?" "Could I test the fading of different kinds of dye on small pieces of cloth?" To know more about these things, you can research and design a whole science fair project about the topic of the durability of different kinds of dye. You will be pleasantly surprised at the number of possible project ideas that will come to mind when you begin to look around and use inquiry questions.

You and those around you each day make an amazing number of statements and ask many questions that could be used to develop science project topics. Be alert and listen for a statement like "She's a clone of her mother." This statement can become an inquiry question, such as "Why do children look like their parents?" or "Why do some children look more like their grandparents than their parents?" These questions could lead you to develop a project about heredity.

Choose a Topic from Your Experience

You may think that you don't have much experience with science topics, but remember that this doesn't have to be rocket science! For example, you have experienced foods changing temperature. You may remember that while you were eating a bowl of ice cream some of the ice cream melted and the melted ice cream tasted sweeter than the still-frozen ice cream. Ask yourself, "Was this because the melted ice cream was warmer?" With research information about receptor cells for taste, you could design a project to determine the effects of temperature on taste. The project problem might be, "How does temperature affect taste?" Propose your hypothesis and start designing your project experiment. For more on developing a project, see Chapter 5, "Project Overview."

Find a Topic in Science Magazines

You can often get ideas for topics in science magazines, but don't expect magazines to include detailed instructions on how to perform experiments and design displays. What you can look for are facts that interest you and that lead you to ask inquiry questions. An article about the Barringer Meteor Crater near Winslow, Arizona, might bring to mind these inquiry questions: "Why don't all meteors cause craters?" "Are Earth's meteor craters like those on the Moon?" Wow! Impact craters is another great project topic you might discover in a science magazine.

Select a Topic from a Book on Science Fair Projects or Science Experiments

Science fair project books, such as this one, can provide you with many different topics to choose from as well as some inquiry questions that can lead to an interesting project. Even though science experiment books do not give you as much direction as science fair project books, many can provide you with exploratory "cookbook" experiments that tell you what to do, what the results should be, and why. But whether you use a science fair project book or a science experiment book, it will be up to you to pick a topic and develop it into a science fair project. The 55 project ideas described in this book can further sharpen your skills at expressing inquiry questions. A list of different project and experiment books can be found in Appendix 8.

Special Topics

Before beginning your project experiment, discuss your plans with your teacher. He or she will be familiar with the regulations that govern potentially dangerous experiments. These may include chemical and equipment usage, experimental techniques, experiments involving live animals, cell cultures, microorganisms, or animal tissues. For some experiments, an adult sponsor trained in the area of your topic will be required to supervise your project. Your safety, as well as the safety of any other people or animals, is the most important thing. In addition, if you have not adhered to the rules of the fair, you may not be allowed to enter your completed work. Before you begin you should have your project topic approved by your teacher. This prevents you from working on an unsafe project and from wasting time on a project that would be disqualified.

Chapter 3

Categories

very science fair has a list of categories, and you need to seek your teacher's advice when deciding which category you should enter your project in. Since science fair judges are required to judge the content of each project based on the category in which it is entered, you would be seriously penalized if you were to enter your project in the wrong category. The list provided here includes the science fair categories represented by the project ideas in Part II with a brief description of each category. Other possible categories not included in Part II include **medicine and health** (study of diseases and health of humans and animals), **engineering** (application of scientific knowledge for practical purposes), **mathematics** (development and application of formal logical systems of computations), and **computer science** (study and development of all aspects of computers, including hardware and software). Ask your teacher for a more comprehensive list of categories and descriptions for the science fair you are entering.

Some topics can correctly be placed in more than one category; for example, the study of soil texture could be in geology or agricultural chemistry. Since you are competing against others, it is to your advantage to select a category that might have fewer competitors. Some representative categories are

* **Astronomy.** The science dealing with all the celestial bodies in the universe, including the planets and their satellites, comets and meteors, the stars and interstellar matter, the star systems known as galaxies, and clusters of galaxies.

  1. **Astrometry.** The observational study concerned with the measurement of the position of celestial bodies.

  2. **Astronautics.** The science and technology of space exploration.

  3. **Astrophysics.** The study concerned particularly with the production and expenditure of radiation in systems such as stars and galaxies and in the universe as a whole.

  4. **Celestial mechanics.** The study of the motion of **celestial bodies** (the natural objects in

  the sky, such as stars, moons, suns, and planets) due to gravity.

  5. **Heliology.** The study of the Sun.

  6. **Planetary satellite science.** The study of moons, the natural satellites of planets. **Lunar** refers to Earth's moon.

  7. **Meteoritics.** The study of meteors.

  8. **Planetology.** The study of the planets of the solar system and especially of how they differ.

* **Biology.** The study of organisms and their life processes.

  1. **Botany.** The study of plants and plant life. Subtopics may include the following:

     a. **Anatomy.** A branch of botany dealing with the structural organization of plants, such as cells and seeds, and the functions of their parts.

     b. **Morphology.** The study of plant forms, development, and life histories. Plant developmental morphology is the study of plant growth and development.

     c. **Physiology.** The study of the physical and chemical processes that take place in living organisms during the performance of life functions, such as photosynthesis and respiration.

  2. **Zoology.** The study of animals and animal life including their structure and growth. Subtopics may include the following:

     a. **Anatomy.** The study of the structure of animals and the function of their body parts.

     b. **Behaviorism.** The study of actions that alter the relationship between an animal and its environment.

  3. **Ecology.** The study of the relationships of living things to other living things and to their environment.

4. **Microbiology.** The study of **microbes** (microscopic living things) or parts of living things.

- **Chemistry.** The study of the composition, structure, properties, and interactions of matter. Subtopics may include the following:

  1. **Agricultural chemistry.** The science that deals with farming concerns, such as the makeup of soil, the application of fertilizer and insecticides, the analysis of agricultural products, and the nutritional needs of farm animals.

  2. **Analytical chemistry.** The science of determining the qualitative and/or quantitative amounts of the chemical components of a substance.

  3. **Biochemistry.** The study of the substances found in living organisms, and of the chemical reactions involved in life processes.

  4. **Inorganic chemistry.** The study of the structure, properties, and reactions of the chemical elements and their compounds (except for hydrocarbons).

  5. **Organic chemistry.** The study of hydrocarbons (compounds composed of carbon and hydrogen) and their reactions.

  6. **Physical chemistry.** The study of the physical properties of substances, such as viscosity and molecular structure.

- **Earth science.** The study of the Earth, from the outermost limits of its atmosphere to the innermost depths of its interior.

  1. **Physical geology.** The study of the nature of Earth in its present state. Subtopics may include the following:

     a. **Hydrology.** The science concerned with the quality and distribution of water on Earth.

     b. **Mineralogy.** The study of the composition and formation of minerals.

     c. **Petrology.** The study of the composition and formation of rocks.

     d. **Soil science.** The study of **soil,** the part of the **regolith** (all the loose rock particles that cover the Earth) that can support rooted plants.

     e. **Topography.** The study of the layout of natural and artificial features on the surface of the Earth.

  2. **Meteorology.** The study of Earth's atmosphere and especially of Earth's weather.

  3. **Oceanography.** The study of oceans and marine organisms. (The study of marine organisms is generally called marine biology.)

- **Physics.** The study of forms of energy and the laws of motion.

  1. **Acoustics.** The study of sound.

  2. **Electrodynamics.** The study of electric and magnetic force fields; the behavior of electrically charged particles in electromagnetic fields; and the propagation of electromagnetic waves.

  3. **Elementary particle physics.** The study of properties of elementary particles (smallest parts of matter) such as electrons in static electricity.

  4. **Fluid dynamics.** The study of the properties and behavior of moving fluids.

  5. **Mechanics.** The study of the forces, interactions, and motions of material.

  6. **Optics.** The study of light.

  7. **Thermodynamics.** The study of temperature and energy; heat flow; the transformation of energy; and phases of matter.

Chapter 4

Project Research

nce you have completed the topic research and selected a topic, you are ready to begin your project research. This research is generally more thorough than topic research. Project research is the process of collecting information from knowledgeable sources, such as books, magazines, software, librarians, teachers, parents, scientists, or other professionals. It also includes data collected from exploratory experimentation. Read widely on the topic you selected so that you understand it and know about the findings of others who have researched the topic. Be sure to give credit where credit is due and record all information and data in your journal.

How successful you are with your project will depend largely on how well you understand your topic. The more you read about your topic and question people who know something about it, the broader your understanding will be. As a result, it will be easier for you to explain your project to other people, especially a science fair judge.

There are two basic kinds of research, primary and secondary.

Primary Research

Primary research is information you collect on your own. This includes information from exploratory experiments you perform, surveys you take, interviews you conduct, and responses to your letters to professionals, specialists, and the like.

One way to obtain information about your topic is to interview people who have special knowledge. These can include teachers, doctors, scientists, or others whose careers require them to know something related to your topic. Let's say your topic is about petrified wood. Who would know about petrification? Start with your science teacher. He or she may have a special interest in fossils or know someone who does. Is there a museum nearby with an exhibit of petrified wood? Is there a web site that provides information about how to contact a scientist knowledgeable about petrification? Owners of rock and mineral shops may have an interest in fossils and

could provide information. You could also try the geology department of a local university. Once you start thinking about your topic, you will discover many people who will serve as valuable resources.

Before contacting the people you want to interview, be prepared. Make a list of questions you want to ask. Try discussing what you know about your topic with someone who knows nothing about it. This can force you to organize your thinking, and you may even discover additional questions to add to your list. Once your list is complete, you are ready to make your call. Simple rules of courtesy, such as the following, will better ensure that the person called will want to help.

1. Identify yourself.

2. Identify the school you attend and your teacher.

3. Briefly explain why you are calling. Include information about your project and explain how the person can help you.

4. Request an interview time that is convenient for the person. This interview could be by telephone, face-to-face, or by e-mail. Be sure to say that the interview will take about 20 or 30 minutes. It may be that the person is free when you call, so be prepared to start the interview.

5. Ask if you may tape-record the interview. You can get more information if you are not trying to write down all the answers.

6. Be on time and be ready to start the interview immediately. Also, be courteous and end the interview on time.

7. If the interview is via e-mail, make sure your questions are understandable and grammatically correct. You might ask someone to edit your e-mail before you send it.

8. At the conclusion of the interview, thank the person for his or her time and for the information he or she provided.

9. Even if you send an e-mail thank you, you may wish to also send a written thank-you note as soon as possible after the interview, so be sure to record the person's name and address.

You may write letters requesting information instead of interviewing, or write letters in addition to interviewing. Check the end of articles in periodicals for lists of names and addresses where you can obtain more information. Your librarian can help you find current periodicals related to your topic. If your project deals with a household product, check the packaging for the address of the manufacturer. Send your letter to the public relations department. Ask for all available printed material about your topic. Send your letter as soon as possible to allow time for material to be sent to you. You can use a form letter similar to the one shown here to make it easier to send your questions to as many different relevant people and organizations as you can find.

Kimberly Bolden
123 Lovers Lane
Travis, TX 00000

October 23, 2003

Rust-Away
223 Iron Street
Metal, NM 11111

Dear Director:

I am a ninth-grade student currently working on a science fair project for the Jenny Lynn High School Science Fair. My project is about conditions affecting the rusting of metals. I would greatly appreciate any information you could send me on the antirusting properties of your product. Please send the information as soon as possible.

Thank you very much.

Sincerely,

Kimberly Bolden

Secondary Research

Secondary research is information and/or data that someone else has collected. You find this type of information in written sources (books, magazines, and newspapers) and in electronic sources (CD-ROM encyclopedias, software packages, or on-line). When you use a secondary source, be sure to note where you got the information for future reference. If you are required to write a report, you will need the following information for a bibliography or to give credit for any quotes or illustrations you use.

Book
Author's name, title of book, place of publication, publisher, copyright date, and pages read or quoted.

Magazine or Periodical
Author's name; title of article; title of magazine; volume, issue number, and date of publication; and page numbers of article.

Newspaper
Author's name, title of article, name of newspaper, date of publication, and section and page numbers.

Encyclopedia
Name of encyclopedia, volume number, title of article, place of publication, publisher, year of publication, and page numbers of article.

CD-ROM Encyclopedia or Software Package
Name of program, version or release number, name of supplier, and place where supplier is located.

On-line Document
Author of document (if known), title of document, name of organization that posted document, place where organization is located, date given on document, and on-line address or mailing address where document is available.

Use Your Research

Now you are ready to use the project research information to express the problem, propose a hypothesis, and design and perform one or more project experiments. The project research will also be useful in writing the project report. The following chapters, 5 through 9, guide you step-by-step through a sample project from start to finish. You may want to read these chapters more than once and refer back to them as you progress through your project.

Chapter 5

Project Overview

efore you start your project, acknowledge that you are preparing an entry for a contest that has rules. One preliminary rule is that your science fair project must represent the work that you do this year. This means that if your project is a continuation of a previous year's project, you must test the variable in a new way or test a new variable.

There are other science fair project rules that must be followed. Your teacher is the source for a list of all of the rules for your fair.

A science fair project is an investigation that uses the scientific method as a tool to discover the answer to some scientific problem. The following information contains a basic outline for using the scientific method in developing a science fair project. While the sample project in this overview is about the Moon, the procedure can be used to assist you in designing a project about any topic. See Chapter 1, "The Scientific Method," for more information about the scientific method.

Research

"Tinkering" research, when you don't know what topic to choose, is the first type of research to be done. Begin by reading different science publications, asking questions of knowledgeable people, and checking out information on the Web. You can also perform exploratory experiments about topics that interest you, such as those listed in Part II of this book. From your research, decide on a topic that you want to discover more information about. (See Chapter 2, for more information about topic research.) For our sample, we'll pick the Moon as our project topic.

Project research, when you have a topic but aren't yet sure what kind of problem you can solve, is the next kind of research to be done. If your topic is the Moon, you would find out as much as possible about this celestial body such as its rates of **rotation** (turning of a body about its **axis**—imaginary line through the center of a body) and **revolution** (one turn of a body about its **orbit**—a curved path about another body). Search astronomy books, periodicals, and the Web for information. As you research, write down

inquiry questions, such as these:

- How do lunar rotation and revolution rates compare? (Lunar is a term that pertains to the Moon.)
- How does Earth's rotation affect lunar motion?

When you're finished writing down questions, select one that most interests you and proceed to the next step.

Project Problem

When you have selected an inquiry question that most interests you, determine whether it can be your science fair problem by asking yourself these questions:

- Is it about animals? If the answer is yes, you will need to check with your teacher about the rules for working with animals.
- Does it compare products? If the answer is yes, check with your teacher to make sure product comparison is an acceptable project. While some local fairs have a special section for product comparisons, others may not allow them.
- Can you state a hypothesis for the question? If the answer is no, then reword the question or select another one.
- Can the question be answered experimentally with measurable results?

How does the sample inquiry question rate as a project problem?

1. "How do lunar rotation and revolution rates compare?" can be determined by reading, but you also could experiment and confirm this for yourself.

2. "How does Earth's rotation affect the motion of the Moon?" can also be determined by reading. If you reword the question as "How does Earth's rotation affect the position of lunar surface features during its apparent daily motion?" you have a question whose answer you could discover for yourself experimentally. This could be done by observing and measuring any angular change in the position

of the points of a crescent-shaped moon each day as it moves from the eastern to the western horizon.

Hypothesis

While the hypothesis is a single statement, it is the key to a successful project. The project experiment will be designed to test the hypothesis, so be sure to propose a hypothesis that is testable with measurable results. For the sample project problem on the effect of Earth's rotation on the position of lunar surface features during the Moon's apparent motion across the sky, an example of a null hypothesis might be, "Earth's rotation has no effect on the position of lunar surface features."

A research hypothesis might be, "If Earth rotates toward the east, then there will be an apparent clockwise rotation of the position of lunar features each day." This is based on the fact that since Earth rotates, the Moon is viewed by observers on Earth from a different direction during the rotation. (See Chapter 1, "The Scientific Method," for more information about research and null hypotheses.)

Project Experiment

Can you think of a way to test your hypothesis experimentally with measurable results? If the answer is no, then you need to go back to the previous step and reword your hypothesis or select another one.

At this stage the project experiment needs to be only a basic design in your mind. A possible experiment for the sample Moon problem might be to observe and photograph a crescent Moon at different times to determine whether there is any rotational change in the lunar features during the Moon's apparent movement from the eastern to the western horizon. This can be done by comparing photographs and the position of the pointed ends of the crescent moon at different times. The control will be the position of the Moon when it is at its **zenith** (highest point) in the sky. Compare the position of the pointed ends of the crescent moons in the photographs taken during an observation time of eight or more hours between moonrise and moonset to determine if any rotational change has occurred. Think about the experiment and ask yourself the following questions. If the answer to any of these questions is no, you need to redesign the experiment.

- Does the experiment have measurable results (results that can be measured with an instrument such as a ruler, scale, stopwatch, or other type of scale, such as angular measurements using your hands as the measuring tool)? For the sample experiment, you could take angular measurements of the position of the crescent points to a line perpendicular to Earth in order to determine whether it appears to rotate each day.
- Does the experiment have an independent variable? For the sample experiment, the length of time of observations is the independent variable.
- Does the experiment have a dependent variable? For the sample experiment, the dependent variable is the angle of the crescent moon.
- Does the experiment have a control? For the sample experiment, the control could be the angle of the crescent moon at its zenith position. The observations recorded before and after this position could be compared to this.
- Does the experiment have constant factors? For the sample experiment, the constant factors include the location of the observer and the type of measuring tool used to make the angular measurements. (See Chapter 1 for more information about variables.)

Get Started!

Once you have a general plan for testing your hypothesis experimentally and recording the data, then you can design the experiment step-by-step and get started. You should be prepared to perform the experiment four or more times, so that you have at least four sets of data. In the sample Moon experiment, you could make observations on four consecutive nights or select four nonsequential nights during a month. For some experiments in which time is a factor, such as one requiring the growth of plants, four or more identical sets of plants and the control could be started at the same time. For all experiments, you should record all the results in your project journal, dating and, if applicable, recording the time of each entry.

Data

Data from your experiment and what you do with it are the main ways that a judge evaluates your experiment. Judges like to see **charts** (data or other information in the form of tables, graphs, or lists) of the measured results. If data are displayed clearly, then judges are likely to conclude that the student understands how to properly develop a science fair project. (For more information about data, see Chapter 6, "Experimental Data.")

Chapter 6

Experimental Data

ata for a science fair project are the quantitative (numerical measurements) observations as well as qualitative (physical characteristics) observations that you obtain experimentally. The information you collect in your project journal is called **raw data.** That simply means data you record during the study. You want your journal to be organized and neat, but you should not recopy the raw data for your journal. Do not be concerned about any stains that might get on the paper while collecting raw data. A clean copy of the raw data can be prepared for your display. As long as you don't change the data, you may represent the information in a different way on your display (such as organized in tables or graphs) so that it is more easily understandable to an observer. (See Chapter 8 for information about the project display.)

Table

Data is generally recorded in a table. In a **table** data is organized into words and numbers in columns and rows. If you picture the table as a grid, the spaces between the vertical lines are called **columns** and the spaces between the horizontal lines are called **rows.** There is no uniform way of designing a table, but all tables should have a title and rows and columns that are labeled. If your table shows measurements, the units of measurement, such as minutes or centimeters, should be part of the column's or row's label. The title generally describes the dependent variable of the experiment, such as "Ants' Attraction to Food," in an experiment where different foods (independent

variable) are used and the number of ants attracted to each food is counted (dependent variable). But the title could express what is being compared, in other words the independent variable versus the dependent variable, "Food Types vs. Attraction of Ants." Your table can also include qualitative notations and drawings to illustrate the data.

Table 6.1, an ant data table, shows the data from an experiment in which the number of ants attracted to foods of different sweetness was examined. In the experiment, four circles of equal diameter were drawn at the same distance from the opening of an anthill and 90° from one another. Three food samples of varying sweetness—very sweet, medium sweet, and unsweetened—were placed in three of the circles. The fourth circle, the control, remained empty. After a period of 5 minutes, the number of ants in each circle was counted and recorded in Table 6.1. The experiment was repeated three times on days with as close to the same constant factors as possible, such as the time of day, temperature, atmospheric conditions (sunny or cloudy), humidity, and so on. In this table the independent variable is the type of food and the dependent variable is the number of ants.

Table 6.2 shows data from a different study in which the problem was to determine if ants communicate by laying a scent trail for other ants to follow. This was done by studying four different anthills. For each anthill, 32 identical pieces of food were placed, eight pieces in each direction—north, south, east, and west—but at equal distances from the anthill opening. A circle of similar diameter was drawn around each food sample. The paths between the food circles and the anthills were called "A," "B," "C," and "D." After 15 minutes, the number of ants found in each circle was counted and part of the ground between the food circles and the anthill on paths A, B, and C were cleaned by rubbing them with a stick. The width of the rubbed section for each path was: A = ½ cm, B = 2 cm, C = 4 cm. The control path, D, was not rubbed. This procedure was done every 15 minutes for a total of 60 minutes. Note that the unit of time, minutes, is part of the column labeled "Time." The independent

TABLE 6.1 ANTS' ATTRACTION TO FOODS DATA					
Type of Food	Number of Ants on Food				
	Test 1	Test 2	Test 3	Test 4	Average
Candy	8	7	9	9	8.25
Fruit	7	7	6	9	7.25
Chips	7	6	7	5	6.25
Control (no food)	3	2	4	4	3.25

variable is the width of the rubbed path and the dependent variable is the number of ants in the food circles. The data for hill number 1 shows the number of ants for paths B and C are about the same and are less than those on paths A and D. It might be concluded that small-width rubbing across path A has little to no affect, thus the total number of ants are about the same as path D, which has no rubbing. A final conclusion would be made by averaging the data from each of the four anthills.

Time, minutes	Number of Ants in Food Circle			
	Path A	Path B	Path C	Path D
0	0	0	0	0
15	9	9	10	8
30	15	8	9	16
45	21	9	7	22
60	32	10	9	33
Total	77	36	35	79

TABLE 6.2 ANT COMMUNICATION DATA—HILL #1

Analyzing and Interpreting Data

When you have finished collecting the data from your experiment, the next step is to **analyze** (critically examine) and **interpret** (restate) the data. Interpreting the data involves reorganizing it into a more easily understood form, such as by graphing it. A **graph** is a diagram used to compare independent and dependent variables. Graphs give a quick visual comparison of data. All graphs should have:

1. A title for the graph.
2. Titles for the *x*-axis (horizontal) and *y*-axis (vertical).
3. Scales with numbers that have the same interval between each division.
4. Labels for categories being counted. Scales often start at zero, but they don't have to.

The three most common graphs used in science fair projects are the bar graph, the line graph, and the circle graph.

In a **bar graph,** solid bars are used to show the relationship between two variables. Bar graphs can have vertical or horizontal bars. The width and separation of each bar should be the same on the graph.

The length of a bar represents a specific number such as 8.25 ants. The width of a bar is not significant and can depend on available space due to number of bars needed. A bar graph has one scale, which can be on the horizontal or vertical axis. This type of graph is most often used when the independent variable is qualitative, such as food types in the ant data in Table 6.1. The independent variable for the Ants' Attraction to Food Data table is food type—candy, fruit, chips, no food (control)—and the dependent variable for this data is the number of ants on each food. A bar graph using the data from Table 6.1 is shown in Figure 6.1. Since the average number of ants from the data varies from 3.25 to 8.25, a scale of 0 to 10 can be used with each unit of the scale representing 0.25 ants. The heights of the bars in the bar graph show clearly that some ants were found on the area of the ground without food, but the greatest number were present in the area where the sweetest food was.

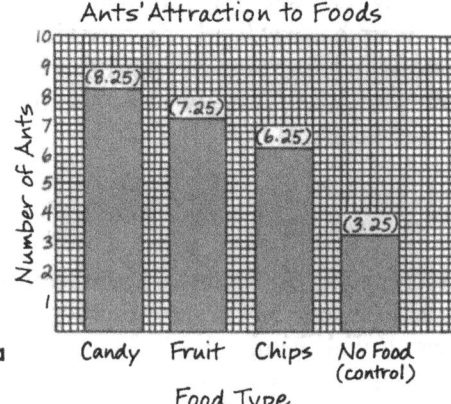

Figure 6.1
Example of a bar graph.

A **line graph** uses a line to compare the relationship between two variables and to show a pattern of change. While a bar graph has one scale, a line graph has two scales. Figure 6.2 shows how the data from Table 6.2 would be arranged on a line graph. Generally the independent variable is on the *x*-axis (horizontal axis) and the dependent variable on the *y*-axis (vertical axis). So for this example, the independent variable of time is on the *x*-axis and the dependent variable of number of ants in path A's food circle is on the *y*-axis. One unit on the time scale represents 1 minute and units are marked off in groups of 15 up to a total of 60 units. One unit on the number of ants scale represents 1 ant. Since the largest number counted was 32 ants, the scale for ants is numbered by 5s from 0 through 35. On the graph, the increase in the slope (incline) of the line over time shows that as time increased more ants were found on the food in path A's food circle.

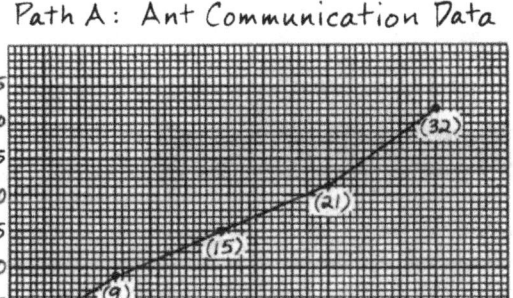

Path A: Ant Communication Data

Figure 6.2 Example of a line graph.

A **circle graph** (also called a **pie chart**) uses a circle divided into wedge-shaped sections to display data.

To plot your data on a circle graph, you need to calculate the size of each section. An entire circle represents 360°, so each section of a circle graph is a fraction of 360°. For example, data from Table 6.1 was used to prepare the circle graph in Figure 6.3. To determine the size of each section in degrees follow these steps:

Ants' Attraction to Foods

8.25 Ants
Candy

119°

7.25 Ants
Fruit

104°

3.25 Ants
No Food
(control)

47°

90°

6.25 Ants
Chips

Figure 6.3 Example
of a circle graph.

1. Express the ratio of each section as a fraction, with the numerator equal to the average number ants counted on each type of food and the denominator equal to the average total number of ants counted on all the food.

Candy = $^{8.25}/_{25}$, Fruit = $^{7.25}/_{25}$, Chips = $^{6.25}/_{25}$, Control = 3.25/25

2. Multiply the fraction by 360°

Candy	8.25/25 × 360° = 119°
Fruit	7.25/25 × 360° = 104°
Chips	6.25/25 × 360° = 90°
Control	3.25/25 × 360° = 47°

To prepare the circle graph, first decide on the diameter needed, then using a compass, draw the circle. Then draw a straight line from the center of the circle to any point on the edge of the circle. Using a protractor, start at this line and mark a dot on the edge of the circle 104° from the line. Draw a line to connect this dot to the center of the circle. The pie-shaped section you made represents the number of ants found on the candy. Start the next section on the line for the candy section. Continue until all the sections are represented. Each section should be labeled, as shown in Figure 6.3.

Each section of a circle graph represents part of the whole, which always equals 100%. The larger the section the greater the percentage of the whole.

To determine the percentage for each section, follow these steps:

1. Change the fractional ratio for each section to a decimal by dividing the numerator by the denominator.

Candy	8.25 ÷ 25 = .33
Fruit	7.25 ÷ 25 = .29
Chips	6.25 ÷ 25 = .25
Control	3.25 ÷ 25 = .13

2. Change the decimal answers to percent. *Percent* means "per hundred." For example, for candy, .33 is read 33/100 or 33 per 100, which can be written as 33%.

Candy	.33 = 33/100 = 33%
Fruit	.29 = 29/100 = 29%
Chips	.25 = 25/100 = 25%
Control	.13 = 13/100 = 13%

You could color each section of the circle graph with a different color and make a **legend** (words written on a graph or diagram to explain it) for the chart as in Figure 6.4. The legend can give the number of ants counted and/or the percentage of the total number of ants.

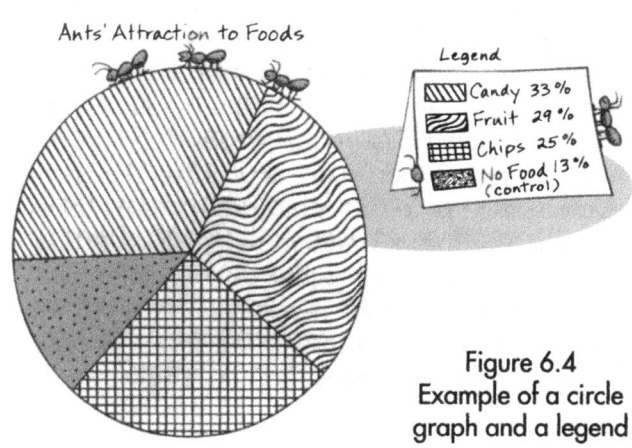

Ants' Attraction to Foods

Legend
Candy 33%
Fruit 29%
Chips 25%
No Food 13%
(control)

Figure 6.4
Example of a circle
graph and a legend

Chapter 7

Project Summaries

ost science fairs require that science fair projects include what are called project summaries. The project summaries include the project journal, an abstract, and a project report. (For information about the project journal, see Chapter 2, "Topic Research.") This chapter gives information on, and examples of, a project abstract and a research paper. Before writing your project summaries, check with your teacher about the requirements for your local science fair as well as those for the International Science and Engineering Fair (ISEF).

Abstract

An abstract is a brief overview of the project. It should be no longer than one page and contain a maximum of 250 words. It includes the title "Abstract," the project title, a statement of purpose, a hypothesis, a brief summary of your experiment procedure, data collected from project experiments, and conclusions. The abstract is generally required to be part of the display. (See Chapter 8 for information about the display.) Contact your teacher for other required uses of the abstract. Most fairs require that a copy of your abstract be displayed on the exhibit. This gives judges something to refer to when making final decisions. It might also be used to prepare an introduction by a special award sponsor. The abstract is a very important representation of your project, so be sure to do a thorough job on this part of your project summary.

The **project title** (a descriptive heading), which appears on your abstract, on the title page of your research paper, and prominently on your display backboard, should capture the theme of the project and be intriguing. Its purpose is to attract the attention of an observer and make him or her want to know more. There are no set rules for the length of the title, but it should be short enough to be read at a glance. A rule of thumb for length is about 10 words or less. A good title for the sample project detailed in Chapter 5 is shown in the sample abstract in Figure 7.1.

Abstract
East to West:
Motion of Lunar Surface Features

[No school or student name should appear on the abstract.]

The purpose of this project is to determine if there is an apparent rotation of the Moon's surface features each day as viewed by an observer from any one specific location on Earth. The hypothesis was that from any specific location on Earth, the Moon is observed from different directions during the day due to Earth's rotation, thus there should be an apparent clockwise rotation of the lunar features each day. The project experiment involved recording the position of the concave side of the crescent moon phase in relation to the Earth's surface as the Moon moved from the eastern to the western horizon. This was done by taking pictures of the Moon at different times before and after it passed its zenith. The pictures were compared to determine any change in position.

The position of the concave side of a crescent phase Moon confirmed the hypothesis that the lunar features on the Moon seen from the Earth apparently rotate in a clockwise direction.

Figure 7.1 Abstract.

Project Report

Your project report is a written report of your entire project from start to finish. The project report should be clear and detailed enough for a reader who is unfamiliar with your project to know exactly what you did, why you did it, what the results were, whether or not the experimental evidence supported your hypothesis, and where you got your research information. This written document is your spokesperson when you are not present to explain your project, but more than that, it documents all your work.

Because you'll be recording everything in your journal as the project progresses, all you need to do in preparing the project report is to organize and neatly copy desired material from the book's contents.

Check with your teacher for the order and content of the report as regulated by the local fair. The following is generally true. The report should be typewritten, double spaced, and bound in a folder or notebook. It should contain a title page, table of contents, introduction, experiment, discussion, conclusion, acknowledgments, and references. The rest of this chapter describes these parts of a project report and gives examples based on the sample project in Chapter 5.

Title Page: This is the first page of the report. The project title should be centered on the page, and your name, school, and grade should appear in the lower right-hand corner.

Table of Contents: This is the second page of your report. The table of contents should contain a list of everything in the report that follows this page, including a page number for the beginning of each section, as shown in Figure 7.2.

Contents

Figure 7.2 Table of Contents.

Introduction: This section sets the stage for your project report. It is a statement of your purpose, along with some of the background information that led you to make this study and what you hoped to achieve from it. It should contain a brief statement of your hypothesis. If it is a research hypothesis, it should state what information or knowledge led you to your hypothesis. If your teacher requires footnotes, then include one for each information source you have used. The sample introduction shown in Figure 7.3 does not use footnotes.

Introduction

The Moon appears to move from east to west each day. Before starting the project, known facts about the Moon included that its revolution and rotation times result in the same side of the Moon always facing Earth. But neither personal knowledge of the effect of Earth's rotation on the motion of the Moon nor observations of the Moon's lunar features during the day had been made.

After reading about the effect of Earth's rotation on the Moon's motion as well as the appearance of lunar features, it was discovered that the Moon's daily east-to-west motion is an apparent motion. It was also discovered by an observer that the Moon's lunar features as viewed from the Northern Hemisphere are rotated clockwise 180° in the Southern Hemisphere. Questions arose about the effect of the direction of the viewer on the apparent rotation of the lunar features.

Curiosity about the upside-down view of the Moon in the Southern Hemisphere resulted in a project that has as its purpose to determine if there is an apparent rotation of the Moon's surface features each day as viewed by an observer from any one specific location on Earth. Based on previously stated research, the hypothesis was that from any specific location on Earth the Moon is observed from different directions during the day due to Earth's rotation, thus there should be an apparent clockwise rotation of the lunar features each day.

Figure 7.3 Introduction.

Experiment: This part of the report contains information about the project experiment. Describe in detail all methods used to collect your data or make your observations. It should include the project problem followed by a list of the materials used and the amount of each, then the procedural steps in outline or paragraph form as shown in Figure 7.4. Note that the experiment described in Figure 7.4 includes instructions for measuring the angle of the Moon's position using photographs. Include instructions for making self-designed equipment or materials in addition to the rest of the experimental procedure. Include any needed photographs and/or drawings of the equipment. All instructions should be written so that anyone could follow them and be expected to get the same results.

Experiment

Problem

How does the Earth's rotation affect the position of lunar surface features as the Moon moves from the eastern to western horizon?

Materials

camera
fine-point permanent black marker
ruler
protractor

Procedure

1. During the **waning** (getting smaller) crescent moon phase, use the camera to take three or more pictures at one- to two-hour intervals of the Moon during its motion along an **arc** (segment of a circle) across the sky from the eastern horizon to its zenith position. Hold the camera so that it is parallel with the ground while taking each photograph and record the time each photo is taken.
2. Repeat step 1 taking pictures of the Moon from its zenith position to the western hemisphere. *Note:* The picture taken at the Moon's zenith will be considered the control.
3. On each photo, use the marker to mark the time the photo was taken.
4. With the marker and ruler, draw a line across the photo that connects the two points of the Moon. Where this line meets the bottom edge of the photo, draw a second line across the photo perpendicular to the bottom edge of the photo.
5. Label the area to the left of the perpendicular line "East" and the area to the right "West".

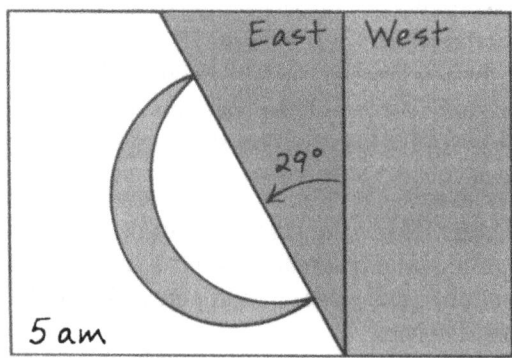

6. Use the protractor to measure the angle between the two lines. Record the measured angle on the photo as well as in a Lunar Feature Angle Data table as shown.
7. Compare the angles of each photo with the control to determine any rotation.
8. Repeat steps 1 through 6 three or more times on consecutive nights or during other Moon phases, such as first and third quarter or **waxing** (growing larger) crescent.

Figure 7.4 Experiment.

Discussion: The discussion of your experimental results is a principal part of your project report. It describes the outcome of your efforts. Include experimental data tables to confirm results, such as Table 7.1. When applicable, determine **random error of measurements** (uncertainty of experimental measurements) and **relative error** (percentage error as compared with a known value). (See Appendix 1 for information about random error of measurement and Appendix 2 for information about relative error.) Include qualitative as well as quantitative results. Never change or omit results because they don't support your hypothesis. Be thorough. You want your readers to see your train of thought, letting them know exactly what you did. Compare your results with published data, commonly held beliefs, and/or expected results. Include a discussion of possible errors. Were your results affected by uncontrolled events? What would you do differently if you repeated this project?

	TABLE 7.1 RESULTS LUNAR FEATURE ANGLE DATA								
	Time								
	5 A.M.	6 A.M.	7 A.M.	8 A.M.	9 A.M.	10 A.M.	11 A.M.	12 P.M.	1 P.M.	2 P.M.
Rotation angle	29°E								

Conclusion: The conclusion is a brief summary, in about one page or less, of what you discovered based on your experimental results, as shown in Figure 7.5. The conclusion states the hypothesis and indicates whether or not the data supports it. If your results are not what you expected, *don't panic*. Assuming that your research led you to your hypothesis, state that while your research backed up your hypothesis, your experimental results did not. Refer to any published data that you based your hypothesis on. Say what you expected and what actually happened. Give reasons why you think the results did not support your original ideas. Include errors you might have made as well as how changing variables other than the independent variable might have affected the results. Discuss any changes you would make to the procedure if you repeated the project and include any ideas for experiments to further investigate the topic of your project. All information in the conclusion should have been reported in other parts of the report so no new material should be introduced in the conclusion.

Acknowledgments: The acknowledgments section is a short paragraph stating the names of people who helped you, with a brief description of their contributions to your project, as shown in Figure 7.6. It should not be just a list of names. Note that when listing relatives, it is generally not necessary to include their names. Identify individuals with their titles, positions, and affiliations (institutions). Note anyone who gave financial support or material donations. Do not reveal monetary amounts of donations.

Conclusion

As stated in my hypothesis, due to the rotation of Earth, there will be an apparent clockwise rotation of the lunar features each day as viewed from any specific location on Earth. The experimental observation over a period of nine hours each day for two days, showed a clockwise rotation of the crescent-shaped Moon, thus a rotation of its lunar features. This data supported my hypothesis and indicated that an observer on Earth sees the moon from a different direction during the day as Earth rotates on its axis.

Figure 7.5 Conclusion.

Acknowledgments

I would like to thank the members of my family who assisted me with this project: my mother, who copy-edited my report, and my father and sister, who assisted in the construction of the display board.

A special note of thanks to Dr. David Russell, professor of astronomy at Calvin University, and to Judith Michael, his assistant, for their expert guidance.

Figure 7.6 Acknowledgments.

References: Your reference list is a bibliography of all the sources where you obtained information. See "Secondary Research" in Chapter 4 for information about bibliographies.

Chapter 8

The Display

our science fair display represents all the work that you have done. It should consist of a backboard and anything else that supports your project, such as models, exhibits or examples, photographs, surveys, and the like. It must tell the story of the project in such a way that it attracts and holds the interest of the viewer. It has to be thorough, but not too crowded, so keep it simple, well organized, attractive, and most of all informative. Your teacher can advise you about which materials that cannot be displayed as well as those that are required, such as your abstract.

The size and shape allowed for the display backboard can vary, so you will have to check the rules for your science fair. Most displays are allowed to be as large as 48 inches (122 cm) wide, 30 inches (76 cm) deep, and 108 inches (274 cm) high (including the table the display stands on). Of course your display may be smaller than this but most participants prefer to take advantage of all the possible space. A three-paneled backboard is usually the best way to display your work. Some office supply stores and most scientific supply companies sell inexpensive backboards. See Appendix 9 for science supply companies from which you can order inexpensive backboards. You can purchase the backboard or build your own. Just remember when building to use materials that are not likely to be damaged during transportation to the fair. Sturdy cardboard or wooden panels can be cut and hinged together.

Purchased backboards generally come in two colors—black and white. You can make them different colors by covering the backboard with self-stick colored shelving paper or cloth. For items placed on the backboard, select colors that stand out but don't distract the viewer from the material being presented. For example, if fluorescent colors are used for lettering and/or background, the bright colors will be what catches the eye instead of your work.

The title and other headings should be neat and large enough to be read at a distance of about 3 feet (1 m). A short title is often eye-catching. For the title and headings, self-stick letters of various sizes and colors, can be purchased at office supply stores and stuck to the backboard. You can cut your own letters

out of construction paper or stencil letters directly onto the backboard. You can also use a computer to print the title and other headings.

Some teachers have rules about the position of the information on the backboard. If your teacher doesn't, put the project title at the top of the center panel, and organize the remaining material in some logical order neatly on the rest of the board. Typical headings used are: Problem, Hypothesis, Experiment (materials and procedure), Data, Results, Conclusion, and Next Time. Figure 8.1 shows one way of placing the material. The heading Next Time, although not always required, would follow the conclusion and contain a brief description of plans for future development of the project. Alternately, this information could be included in the conclusion rather than under a separate heading.

Typed material can be secured to colored backing such as construction paper. Leave a border of about ¼ to ½ inch (0.63 to 1.25 cm) around the edges of each piece of typed material. Use a paper cutter to cut papers so that the edges will be straight.

You want a display that the judges will remember positively. So before you glue everything down, lay the backboard on a flat surface and arrange the materials a few different ways. This will help you decide on the most suitable and attractive presentation. Figure 8.1 shows what a good display might look like.

Helpful Hints

1. Make sure the display represents the current year's work only.

2. The title should attract the interest of a casual observer. Not only should the title itself be interesting but it should stand out visually.

3. Organization is a very important part of designing a display. You want a logical order so that observers (especially judges) can easily follow the development of your project from start to finish. Before you actually stick anything to the board, make a diagram of where each part will be placed.

Figure 8.1 Example of a good display.

4. One way to arrange the letters on the backboard is to first lay the letters out on the board without attaching them. Then use a yardstick (meterstick) and pencil to draw a straight, light guideline where the bottom of each letter should line up. This will help you keep the lettering straight. Before adhering everything, you may wish to seek the opinion of other students, teachers, or family members.

5. If you need electricity for your project, be sure the wiring meets all safety standards.

6. Bring an emergency kit to the science fair that includes anything you think you might need to make last-minute repairs to the display, such as extra letters, glue, tape, construction paper the color of the backboard, a stapler, scissors, pencils, pens, touch-up paint, markers, and so forth.

7. If allowed, before standing your backboard on the display table, cover the table with a colored cloth. Choose a color that matches the color scheme of the backboard. This will help to separate your project from other projects displayed next to yours.

Do's and Don'ts

Do adhere to the size limitations and safety rules set by the fair you are entering. Generally the size limitations are 30 inches (76 cm) deep, 48 inches (122 cm) wide, and 72 inches (183 cm) high (does not include table height).

Do use computer-generated graphs or neatly hand-drawn ones.

Do display photos representing the procedure and the results.

Do use contrasting colors.

Do limit the number of colors used and do not use fluorescent colors.

Do display models when applicable. If possible, make the models match the color scheme of the backboard.

Do attach charts neatly. If there are many, place them on top of each other so that the top chart can be lifted to reveal the ones underneath.

Do balance the arrangement of materials on the backboard. This means to evenly distribute the materials on the board so that they cover about the same amount of space on each panel.

Do use adhesives, such as a glue stick, rubber cement, or double-sided tape to attach papers. Liquid glue generally causes the paper to wrinkle.

Do check spelling, math calculations, and chemical formulas.

Do have one or more adults, such as teachers and/or parents, review all of your work before you put it on the backboard.

Don't leave large empty spaces on the backboard.

Figure 8.2 Example of a bad display.

Don't leave the table in front of the backboard empty. Display your models (if any), report, copies of your abstract, and your journal here.

Don't hang electrical equipment on the backboard so that the electric cord runs down the front of the backboard.

Don't make the title or headings hard to read by using uneven lettering or words with letters of different colors.

Don't hand print the letters on the backboard.

Don't attach folders that fall open on the backboard.

(Figure 8.2 shows how *not* to set up your display.)

Safety

Anything that is or could be hazardous to other students or the public is *prohibited* and cannot be displayed. The following is a list of things that are generally unacceptable for display. Your teacher has access to a complete list of safety rules from your local science fair officials. Models or photographs can be used instead of things that are restricted from display.

Unacceptable for Display

1. Live animals

2. Microbial cultures or fungi, living or dead

3. Animal or human body parts, except teeth, hair, nails, and dried animal bones

4. Liquids, including water

5. Chemicals and/or their empty containers, including caustics, acids, and household cleaners

6. Open or concealed flames

7. Batteries with open-top cells

8. Combustible materials

9. Aerosol cans of household solvents

10. Controlled substances, poisons, or drugs

11. Any equipment or device that would be hazardous to the public

12. Sharp items such as syringes, knives, or needles

13. Gases

Chapter 9

Presentation and Evaluation

our teacher may require that you give an oral presentation on your project for your class. Make it short but complete. Presenting in front of your classmates may be the hardest part of the project. You want to do your best, so prepare and practice, practice, practice. If possible, tape your practice presentation on a tape recorder or have someone videotape you. Review the tape and/or video and evaluate yourself, then review your notes and practice again.

Practicing an oral presentation will also be helpful for the science fair itself. The judges give points for how clearly you discuss the project and explain its purpose, procedure, results, and conclusion. Judges are impressed with students who can speak confidently about their work. They are not interested in memorized speeches—they will want to have a conversation with you to determine if you understand the work you have done from its start to its finish. While the display should be organized so that it explains everything, your ability to discuss your project and answer the judges' questions convinces them that you did the work and understand what you have done. Practice a speech in front of friends, and invite them to ask questions. If you do not know the answer to a question, never guess or make up an answer or just say "I don't know." Instead, say that you did not discover that answer during your research, and then offer other information that you found of interest about the project. Be proud of the project, and approach the judges with enthusiasm about your work.

As you progress through developing your project, keep in mind that you may be asked about different developmental stages. Take note of some of the ideas that you had while working on your project. These can be used to hold an audience's interest and impress judges.

You can decide on how best to dress for a class presentation, but for the local fair it is wise to make a special effort to look nice. You are representing your work. In effect, you are acting as a salesperson for your project, and you want to present the very best image possible. Your appearance shows how much pride you have in yourself, and that is the first step in introducing your product, your science project.

Judging Information

Most fairs have similar point systems for judging science fair projects. You may be better prepared if you understand that a judge generally starts by assuming that each student's project is average. Then he or she adds or subtracts points from that level. A student should receive points for accomplishing the following.

1. Project Objectives
   - Presenting original ideas
   - Stating the problem clearly
   - Defining the variables and using controls
   - Relating background reading to the problem

2. Project Skills
   - Being knowledgeable about equipment used
   - Performing the experiments with little or no assistance except as required for safety
   - Demonstrating the skills required to do all the work necessary to obtain the data reported

3. Data Collection
   - Using a journal to collect data and research results
   - Repeating the experiment to verify the results
   - Spending an appropriate amount of time to complete the project
   - Having measurable results

4. Data Interpretation
   - Using tables, graphs, and illustrations in interpreting data

- Using research to interpret data collected

- Collecting enough data to draw a conclusion

- Using only data collected to draw a conclusion

5. Project Presentation (Written Materials/Interview/ Display)

- Having a complete and comprehensive report

- Answering questions accurately

- Using the display during the oral presentation

- Justifying conclusions on the basis of experimental data

- Summarizing what was learned

- Presenting a display that shows creative ability and originality

- Presenting an attractive and interesting display

If the project lacks any of these things, then points are taken away.

Do's and Don'ts at the Fair

Do bring activities, such as puzzles to work on or a book to read, to keep yourself occupied at your booth. There may be a lengthy wait before the first judge arrives, and even between judges.

Do become acquainted with your neighboring presenters. Be friendly and courteous.

Do ask neighboring presenters about their projects, and tell them about yours if they express interest. These conversations pass time and help relieve nervous tension that can build when you are waiting to be evaluated. You may also discover techniques for research that you can use for next year's project.

Do have fun!

Don't loudly laugh or talk with your neighbor.

Don't forget that you are an ambassador for your school. This means that your attitude and behavior influence how people at the fair think about you and the other students at your school.

II

55 A+
SCIENCE FAIR
PROJECT IDEAS

Astronomy

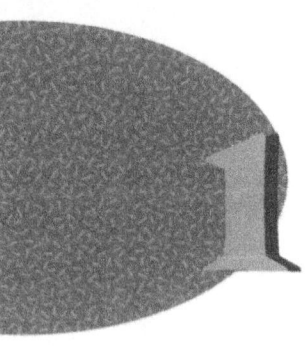

Solar System Scale
Miniature Model

Our solar system consists basically of the Sun, the nine planets and their moons, and space debris. The boundary of the solar system is called the heliopause, which is the limit the solar winds reach. Solar winds are streams of charges escaping from the Sun's atmosphere—gaseous layer surrounding it—and flowing into the solar system. From the heliopause, the solar system is about 18.6 billion miles (29.8 billion km) across. The Sun is the largest celestial body in the solar system, having a diameter of about 870,000 miles (1,392,000 km). Pluto is the smallest planet at about 1,434 miles (2,294 km) across. All of the millions of asteroids and comets are much smaller than Pluto.

In this project, you will make a 3-D model to represent the sizes of celestial bodies in the solar system. You will also learn how to represent the distances between planets and between planets and their moons.

Getting Started

Purpose: To build a 3-D scale model of Earth.

Materials

drawing compass
metric ruler
sheet of colored poster board
scissors
index card
pen

Procedure

1. The diameter of Earth is about 7,973 miles (12,757 km). Using the metric scale of 1 cm/3,000 km, calculate the diameter of the model Earth as follows.

    Earth's actual diameter ÷ 3,000 km/cm

    = model Earth's diameter

    12,757 km ÷ 3,000 km/cm = 4.252 cm

    Rounding the number to the closest centimeter, the model Earth's diameter would be 4 cm.

2. Use the compass to draw two circles with diameters of 4 cm (radius 2 cm) on the poster board.

3. Cut out the circles, then cut along a straight line (radius) from the circumference to the center of each circle.

4. Fit the two circles together at a 90° angle to each other (see Figure 1.1).

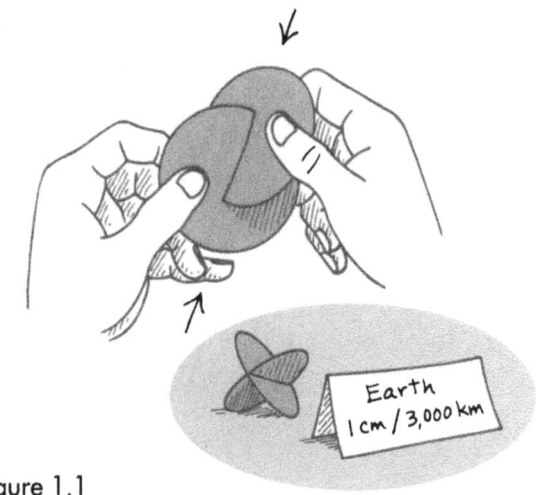

Figure 1.1

5. Prepare a legend showing the scale of the model by folding the index card in half. Write the scale, "1 cm/3,000 km," on one side of the card. Stand the card alongside the model.

Results

You have made a scale model of Earth.

Why?

Celestial bodies are groups of natural objects in the sky, such as suns, stars, planets, and moons. A **solar system** is a group of celestial bodies that **revolves** around the sun. This means they move in a curved path around another object, which is also called **orbiting**. A **sun** is a **star** (glowing sphere of gas) that is the celestial body at the center of a solar system. If capitalized, Sun is the central body in our own solar system. Basically our solar system is made up of nine **planets** (celestial bodies that orbit a sun and shine only by reflected light), **moons** (small bodies in orbit around a planet) and other space **debris** (scattered pieces of something that has been broken). If capitalized, Moon refers to Earth's moon.

In this experiment you made a **scale model** (replica made in proportion to the object it represents) of the Earth, one of the planets of our solar system. A **scale** is a ratio between the measurements of a model and the actual measurements of an object. In a scale ratio the model measurement comes first. If the first value of the ratio is smaller, the scale drawing or model is reduced, such as the 1 cm/3,000 km in this experiment. Scale models of bodies in our solar system, such as Earth, are valuable because they allow an observer to compare the relative sizes and distances of large things more easily.

Try New Approaches

How does the size of Earth compare with the other planets in the solar system? Make more scale models, using the diameters of the planets given in Appendix 3. **Science Fair Hint:** Attach threads to the models and suspend them from a 4-foot (1.2-m) or longer dowel. Support the ends of the dowel, then take a picture of the models. Use the photo and models in your science fair display.

Design Your Own Experiment

1. A **satellite** is a body that revolves about a celestial body. A **natural satellite** is a celestial body that revolves about a larger celestial body. Earth has one natural satellite, the Moon. Design a scale model that accurately represents the size of both Earth and the Moon and the average distance between them. The Moon's diameter is 2,173 miles (3,476 km). The average distance between the centers of Earth and the Moon is about 240,250 miles (384,400 km). Use the scale 1 cm/3,000 km. Make one sphere to scale for Earth and another for the Moon. Label the models "Earth" and "Moon." Calculate the length of string needed to represent the distance between them. Measure and cut string of the necessary length. Tape one end of the string to the

center of the model Moon and the other end to the center of the model Earth. Ask helpers to hold the models, stretching the string taut between the models, while you take a photo (see Figure 1.2). Display the photo along with a legend indicating the actual sizes and distances along with the scale you used.

2. The planet Jupiter has many moons. Its four largest are collectively called the **Galilean satellites** (any of the four largest moons of Jupiter: Io, Europa, Ganymede, and Callisto) because they were discovered in 1610 by Galileo (1564–1642). Make a model of Jupiter and the Galilean satellite called Ganymede. The diameter of this moon is 3,293 miles (5,268 km). It orbits 668,750 miles (1,070,000 km) from Jupiter. Jupiter's diameter is 89,875 miles (143,800 km).

3a. The average distance between Earth and the Sun is about 93 million miles (149 million km). This distance is called an **astronomical unit** (AU) and is used as a measure of distance in the solar system. The AU measurement for any planet can be calculated by dividing the planet's distance in miles (km) by 93 million miles (149 million km). For example, Mercury is about 36 million miles (58 million km) from the Sun, thus the AU distance between the Sun and Mercury is:

Figure 1.2

33

36,000,000 miles ÷ 93,000,000 miles = 0.387 AU

58,000,000 km ÷ 149,000,000 km = 0.389 AU

Rounding the answer to the nearest tenth, Mercury is 0.4 AU from the Sun.

Prepare a data table like Table 1.1, giving the average distances of the planets in miles (km) from the Sun (distances shown are from Appendix 3) and in AU units. Calculate the AU distance to the nearest tenth as in the preceding example.

TABLE 1.1 PLANET'S AVERAGE DISTANCES FROM THE SUN		
Planet	Average Distance, millions of miles (millions of km)	Distance, AU
Mercury	36 (58)	0.4
Venus	68 (108)	
Earth	93 (149)	1.0
Mars	143 (228)	
Jupiter	486 (778)	
Saturn	892 (1,427)	
Uranus	1,794 (2,870)	
Neptune	2,810 (4,497)	
Pluto	3,688 (5,900)	

b. Use the AU distances to draw a scale model of the solar system.

Get the Facts

1. By definition, natural satellites (moons) are smaller than the planets they orbit. Even the smallest known planet in our solar system, Pluto, has a moon called Charion. Pluto is smaller than Earth's moon, and Charion is even smaller. Find out more about the natural satellites of the planets. Do all the planets have satellites? How do their sizes compare with the size of the planets they orbit? Make models of the planets and their satellites. For information, see Patrick Moore's and Will Tirion's, *Guide to Stars and Planets* (Cambridge: Cambridge University Press, 1993).

2. The sixth planet from the Sun, Saturn, is known for particles that orbit the planet, creating what looks like rings as observed from Earth. Make a model of Saturn. How many rings are there? What are they made of? How large are they? What causes them to change in visibility? For information, see Ann-Jeanette Campbell's *New York Public Library Amazing Space* (New York: Wiley, 1997), pp. 124–128.

Notes on
Solar System Scale

Key Facts:

_____
_____
_____
_____

My Results:

_____
_____
_____
_____
_____
_____
_____
_____

Conclusions:

_____
_____
_____
_____
_____

Results of Try New Approaches:

_____
_____
_____
_____
_____

Notes on Designing My Own Experiment:

_____
_____
_____
_____
_____
_____
_____
_____
_____
_____
_____
_____
_____
_____
_____
_____
_____
_____
_____
_____
_____
_____

Notes on Get the Facts:

_____
_____
_____
_____
_____
_____

2 Barycenter
The Balancing Point

Our solar system consists of the Sun and the many millions of celestial bodies, including large planets and microscopic dust particles, which orbit around it. As a unit, the solar system has a center of mass, its balancing point. At this point, the system would balance like a spinning plate atop a circus performer's balancing stick. This point, called the barycenter, is the exact point about which all the bodies in the solar system orbit. Since the Sun is vastly larger and heavier than all the other bodies combined, the solar system's barycenter is very close to the Sun—but not at the Sun's center. Thus, while all the other solar system bodies seem to orbit the Sun, they, including the Sun, are actually orbiting a point in space just beyond the Sun's outer layer.

Binary bodies are two celestial bodies held together by mutual gravitational attraction. In this project, you will learn how mass affects the location of their barycenter. You will discover the mathematical relationship between the masses of binary bodies and their distances from their barycenter. You will prepare a model of the Earth-Moon system and determine the orbit that each body follows as it orbits the barycenter of the system. You will also discover the location of the barycenter for most planet-moon systems.

Getting Started

Purpose: To model the barycenter of binary bodies.

Materials

one-hole paper punch
½-by-3-inch (1.25-by-7.5-cm) piece of thick paper, such as a file folder
6-foot (1.8-m) or longer cord
1 pound (454 g) modeling clay
food scale
⁵⁄₁₆-by-48-inch (0.78-by-120-cm) wooden dowel
yardstick (meterstick)

Procedure

1. Use the paper punch to make a hole in each end of the piece of paper.

2. Bend the paper to bring the holes together. Thread one end of the cord through the holes. Tie a knot to hold the holes together. You have made a paper sling for the dowel.

3. Tie the other end of the cord to a ceiling hook or other supporting object. Adjust the length of the cord so that the paper sling hangs about chest high.

4. Using the food scale, measure two 8-ounce (227-g) pieces. Shape each piece into a ball.

5. Stick one end of the dowel into one of the clay balls to a depth equal to the radius of the ball.

6. Slide the free end of the dowel through the paper sling.

7. Repeat step 5 using the remaining clay ball on the other end of the dowel.

8. Determine the balancing point by moving the dowel back and forth in the sling until it balances (see Figure 2.1).

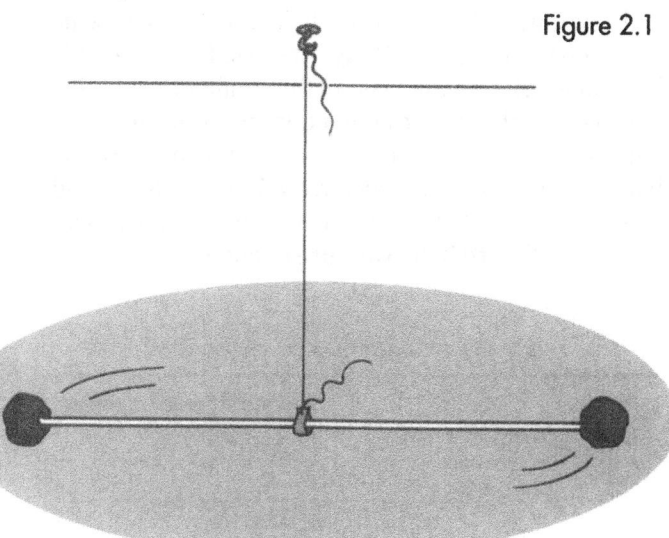

Figure 2.1

9. Measure and compare the distance between the center of each clay ball and the center of the paper sling.

10. Gently push one of the balls so that the dowel turns. Observe the motion of the clay balls.

Results

The balancing point is in the center of the dowel, an equal distance from each of the clay balls. The balls move in a circular path around the sling.

Why?

Binary bodies are two celestial bodies held together by mutual gravitational attraction. **Gravity** is a force of attraction between all objects in the universe. Examples of binary bodies are two stars, a planet and its sun, or a planet and its moon. Binary bodies behave somewhat as if they were connected by a dowel. Their center of gravity is called the **barycenter** (the point between two binary bodies where their mass seems to be concentrated and the point about which they rotate). If the masses of the binary bodies are equal, the barycenter lies at an equal distance from each body. Binary bodies **revolve** (move in a circular path about a point) about their barycenter.

Try New Approaches

Different masses affect the distance of each body from the barycenter. Mold the clay into one large ball and one tiny ball. Weigh the balls on the food scale and prepare a Celestial Body Mass vs. Distance table like Table 2.1. Call the mass of the smaller ball m_1. Let the mass of the larger body equal m_2. Call their distances from the center of each clay ball to the barycenter d_1 and d_2, respectively. Use the dowel and sling to find the barycenter as in the original experiment (see Figure 2.2). Measure d_1 and d_2.

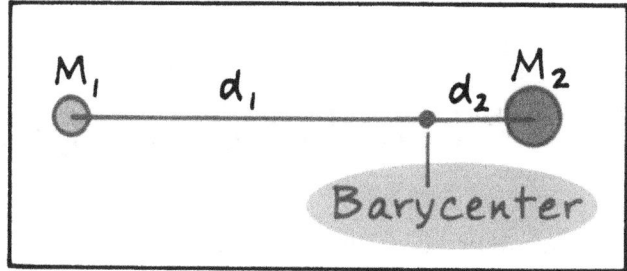

Figure 2.2

For example, when m_1 = 1 ounce (28 g) and m_2 = 15 ounces (426 g), and d_1 = 45 inches (112.5 cm), then d_2 = 3 inches (7.5 cm). Notice that mass and density have an inverse relationship, which means that when one term increases, the other decreases.

In our example, $m_1 \times d_1 = m_2 \times d_2$, or $m_1/m_2 = d_2/d_1$ = 3/45, which reduces to 1/15. Thus, m_1 lies 15 times farther from the barycenter than does m_2, and m_2 is 15 times more massive than m_1.

Design Your Own Experiment

The Earth and Moon are binary bodies with a barycenter that lies about 1,000 miles (1,600 km) beneath Earth's surface on the side facing the Moon. Design a model to show the paths of Earth and the Moon as they orbit their barycenter. One way is to use a 2-by-12-inch (5-by-30-cm) strip of poster board. In the center of one end of the strip make a dot and draw a circle with a 4-mm radius around the dot. Label this circle "Earth." Lay a ruler across the length of the poster board and make two dots at these distances from the center of the circle: 3 mm and 240 mm. Label the first dot "Barycenter." Draw a second circle as small as you can around the second dot. Label this circle "Moon." Stick a pushpin through the barycenter dot into a pencil eraser on the underside of the dot (see Figure 2.3). Holding the pencil in one hand, rotate the strip around the pushpin in a counterclockwise direction. Use the equation in the previous experiment and these distances to calculate the ratio of the Earth/Moon masses. Compare this ratio with a ratio of Earth's and the Moon's mass found in a reference book. Display the model and a legend of the scale.

TABLE 2.1 CELESTIAL BODY MASS VS. DISTANCE		
Celestial Body	Mass	Distance from Barycenter, inches (cm)
1	m_1 = ?	d_1 = ?
2	m_2 = ?	d_2 = ?

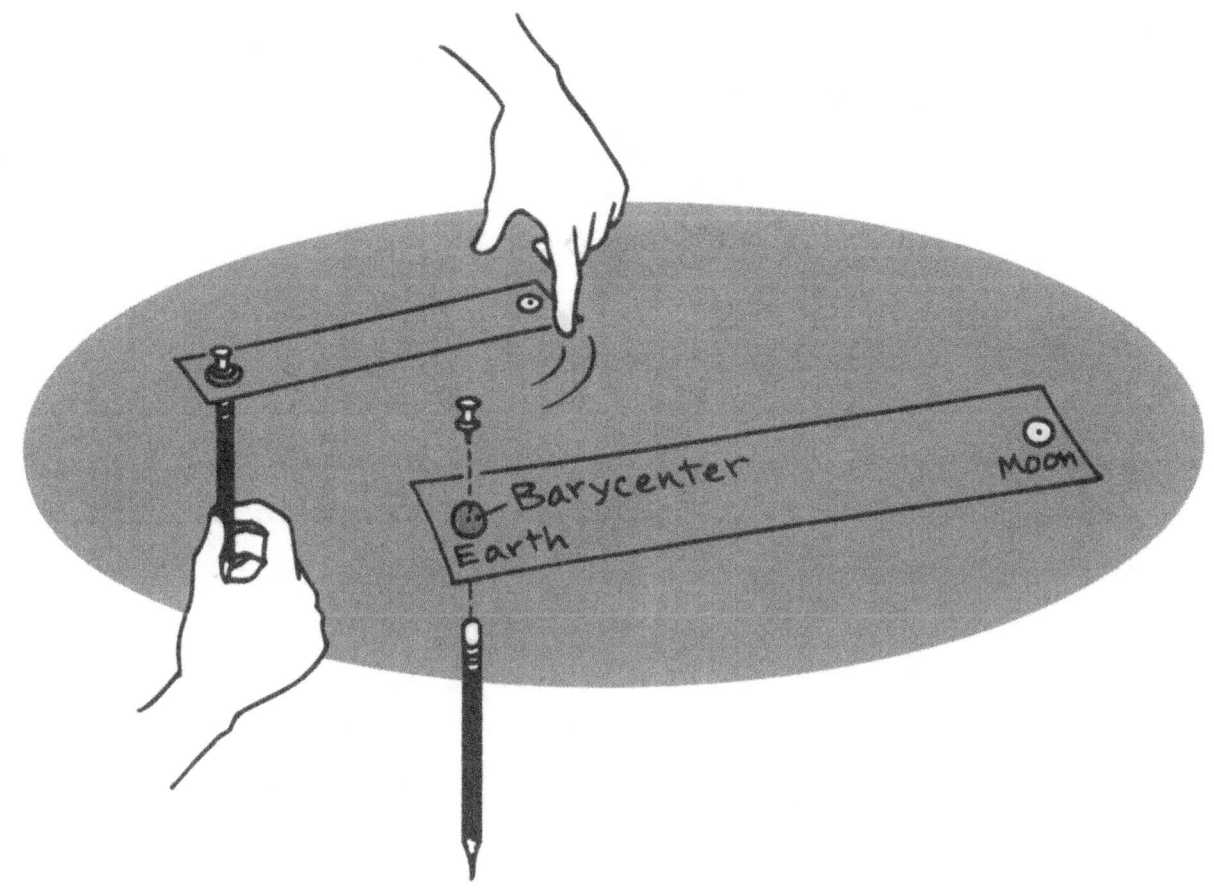

Figure 2.3

Get the Facts

The average barycenter of our solar system lies just outside the surface of the Sun. It changes depending on the location of the planets. Jupiter, the most massive planet, has the greatest effect. Find out more about the barycenter of binary bodies in the solar system. Where does the barycenter lie for most planet-satellite (moon) systems? Which planet has such a massive moon that the barycenter lies in the space between them? For information, see Thomas R. Watters, *Planets: A Smithsonian Guide* (New York: Macmillan, 1995).

NAME

Notes on
Barycenter

Key Facts:

_____
_____
_____
_____

My Results:

_____
_____
_____
_____
_____
_____
_____
_____

Conclusions:

_____
_____
_____
_____
_____

Results of Try New Approaches:

_____
_____
_____
_____
_____

Notes on Designing My Own Experiment:

_____
_____
_____
_____
_____
_____
_____
_____
_____
_____
_____
_____
_____
_____
_____
_____
_____
_____
_____
_____
_____
_____
_____
_____

Notes on Get the Facts:

_____
_____
_____
_____
_____
_____

3 Orbital Eccentricity
How Circular a Celestial Body's Path Is

Planets are large celestial bodies that orbit a sun. In our solar system, the nine known planets are Mercury, Venus, Earth, Mars, Jupiter, Saturn, Uranus, Neptune, and Pluto. The seventeenth-century German astronomer Johannes Kepler (1571–1630) discovered that planets follow an elliptical path. A planet's orbital eccentricity is how much the orbit deviates from being a circle.

In this project, you will discover the basic shape of planetary orbits as stated in Kepler's first law of planetary motion. You will also determine how nearly circular orbits are and make models.

Getting Started

Purpose: To draw the planets' orbital shape as described by Kepler.

Materials

pencil
ruler
12-inch (30-cm)–square piece of white poster board
12-inch (30-cm)–square piece of thick cardboard
2 pushpins
10-inch (25-cm) piece of string

Procedure

1. Draw a straight line across the center of the poster board.

2. Near the center of the line, draw two dots 4 inches (10 cm) apart.

3. Place the poster board on the cardboard. Stick one pushpin in each of the dots on the line.

4. Double the string and tie a knot as close to the free ends as possible.

5. Position the loop of string around the pushpins.

6. Place the pencil point against the inside of the loop.

7. Keeping the string taut, guide the pencil around the inside of the string to draw a closed curved on the paper (see Figure 3.1). Keep the paper and loop of string for the next experiment.

Figure 3.1

Results

You have drawn a closed curve with a longer, major axis (greatest distance across) and a shorter, minor axis (least distance across), as shown in Figure 3.2.

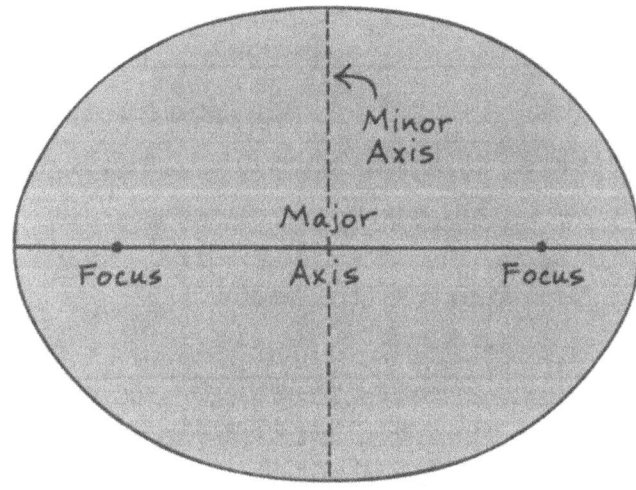

Figure 3.2

Why?

The closed curve in this investigation is an oblong figure called an ellipse. An ellipse is the figure drawn around two fixed points called the **foci** (the singular is **focus**). The distance from one focus to any point on the ellipse and back to the other focus is equal to the length of the major axis. In 1609, Kepler published his **first law of planetary motion,** which states that the orbits of the planets around the Sun are ellipses in which the Sun is at one focus.

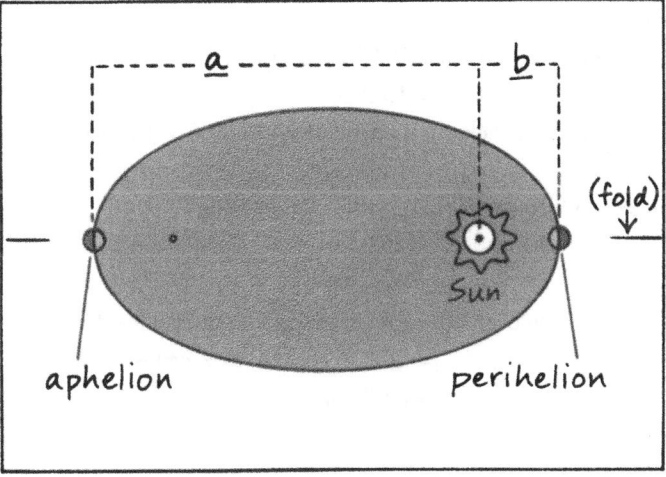

Figure 3.3

Try New Approaches

1. How does the distance between the foci affect the shape of the orbit? Draw four circles with the pushpins these distances apart: 0 inches (0 cm), 3 inches (7.5 cm), 5 inches (12.5 cm), and the length of the loop of string.

2. **Eccentricity** is a ratio that describes how much an ellipse deviates from the shape of a perfect circle. To calculate eccentricity, divide the distance between the foci by the length of the major axis. Calculate the eccentricity of each of the orbits you have drawn. Use the results to formulate statements comparing the eccentricity values of different ellipses. For example, as the eccentricity increases, the shape is more or less circular/ elliptical/linear.

Design Your Own Experiment

1a. Aphelion is the point in a planet's orbit farthest from the Sun. Perihelion is the orbital point closest to the Sun. Design an investigation to use the distance of a planet at aphelion and at perihelion to determine an orbit's eccentricity. One way is to fold a sheet of white copy paper in half, long sides together. Open the paper and stick two pushpins on the fold line at a distance apart less than the length of a loop of string. Place the loop around the pushpins and draw an ellipse as before. Draw a sun at one focus and two small circles of equal size on the ellipse, one at each end of the major axis (see Figure 3.3). The

circles represent a planet at perihelion and aphelion. Measure the perihelion distance (*a*) and aphelion distance (*b*) on your drawing. Use your measurements and the following example to determine the eccentricity of the planet.

$$a = \text{distance at aphelion}$$
$$b = \text{distance at perihelion}$$
$$a - b = \text{distance between foci}$$
$$a + b = \text{length of major axis}$$
$$e = \text{eccentricity}$$
$$e = (a - b) \div (a + b)$$
$$= ?$$

b. Calculate the eccentricity of each of the nine planets using their perihelion and aphelion distances in Appendix 3. Display your work in a table that includes columns for the following data on each planet: distance at aphelion and at perihelion, distance between foci, length of major axis, and eccentricity.

2. Make scale drawings of the orbits of Mercury, Venus, Earth, and Mars. Let 1 cm equal 10 million km (1.0×10^7). To determine where to place the pushpins, you need to determine the focal difference using the perihelion and the aphelion, and the length of the major axis. The length of the string will be twice the sum of the aphelion plus the perihelion, plus about 4 cm to tie the knot in the string. Prepare a measurement table similar to Table 3.1, then use the table to make your drawings.

Example:

$$\text{Earth's aphelion} = (1.52 \times 10^8 \text{ km}) \div (1.0 \times 10^7 \text{ km/cm})$$
$$= 15.2 \text{ cm}$$
$$\text{Earth's perihelion} = (1.47 \times 10^8 \text{ km}), (1.0 \times 10^7 \text{ km/cm})$$
$$= 14.7 \text{ cm}$$
$$\text{distance between foci} = \text{aphelion} - \text{perihelion}$$
$$= 15.2 \text{ cm} - 14.7 \text{ cm}$$
$$= 0.5 \text{ cm}$$
$$\text{length of string} = 2 \times (\text{aphelion} + \text{perihelion}) + 4 \text{ cm}$$
$$= 2 \times (15.2 \text{ cm} + 14.7 \text{ cm}) + 4 \text{ cm}$$
$$= 63.8 \text{ cm}$$

Get the Facts

Johannes Kepler was an excellent mathematician who used other scientists' observations to develop a theory of planetary motion. On whose work did Kepler rely? How long did he work to arrive at his explanations? How did Kepler's model compare with the models of earlier astronomers, such as Aristotle, Ptolemy, and Copernicus? For information, see David Filkin, *Stephen Hawking's Universe* (New York: Basic Books, 1997), pp. 33–38.

TABLE 3.1 SCALE OF ORBITAL DISTANCES

Planet	Aphelion, cm	Perihelion, cm	Distance between Foci, cm	String length, cm
Earth	15.2	14.7	0.5	63.8

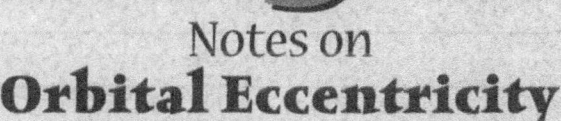

Notes on
Orbital Eccentricity

Key Facts:

_____

_____

_____

_____

My Results:

_____

_____

_____

_____

_____

_____

_____

Conclusions:

_____

_____

_____

_____

Results of Try New Approaches:

_____

_____

_____

_____

Notes on Designing My Own Experiment:

Notes on Get the Facts:

4 Angular Separation
Angular Distance between Celestial Bodies

The apparent distance between celestial bodies is how large the linear measurement between bodies appears to be from Earth. Angular separation or angular distance is the apparent distance expressed in radians or degrees.

In this project, you will build a cross-staff resembling the instrument used by early navigators and astronomers. You will use it to measure both angular separation and angular diameter of celestial bodies including the Moon.

Getting Started

Purpose: To use a cross-staff to measure angular separation.

Materials

cross-staff (see Appendix 5)

Procedure

1. Use the cross-staff to measure angular separation. You can take measurements inside or outdoors.

   • Stand 4 yards (4 m) from a closed door or other object. Record this sighting distance in column 1 of an Angular Separation (Wide Sight) Data table like Table 4.1.

   • Rest the zero end of the measuring stick against one of your cheekbones. Open the eye over this cheekbone and close the other eye. Use your open eye to sight along the length of the stick. With the bottom of the crosspiece

parallel with the floor, slide the crosspiece until the left side of the door lines up with the left-hand edge of the wide side and the right side of the door lines up with the right-hand edge of the wide sight (see Figure 4.1). The width of the wide sight is d_1 and is equal to 4 inches (10 cm). Record this sight width in column 2 of your data table as shown.

Figure 4.1

   • Read the value on the measuring stick where the bottom of the labeled side of the crosspiece touches the measuring stick. This measurement is d_2, the distance from the sight to your eye. Record measurement d_2 in column 3 of your data table for trial 1.

Sighting Distance	Distance: Width of Sight (d_1), inches (cm)	Distance: Sight to Eye (d_2) inches (cm)						Angle Separation Da,°	Random Error of Measurement
		Trial 1	Trial 2	Trial 3	Trial 4	Trial 5	Average		
4 yards (4 m)	4 inches (10 cm)								

TABLE 4.1 ANGULAR SEPARATION (WIDE SIGHT) DATA

2. Repeat step 1 four times, for a total of five independent measurements of the same distance.

3. Calculate the angular separation between the left and right sides of the door in degrees, 0°, using the average of your five measurements for d_2 and this equation:

$$D_a = 57.3° \times (d_1 \div d_2)$$

where D_a is the angular separation, d_1 is the width of the sight, and d_2 is the distance from the sight to your eye. D_a is expressed in degrees. Both d_1 and d_2 must be expressed in the same unit—either inches or centimeters. *Note: $d_1 \div d_2$ yields a number without a unit of measurement.* When no unit of measurement is indicated in giving the measure of an angle, the angle is understood to be expressed in radians. To express the angle in degrees, the conversion 57.3° per 1 radian is used.

For example, for the wide sight, if d_1 = 4 inches (10 cm) and d_2 = 20 inches (50 cm), then

$$D_a = 57.3° \times d \,(4 \text{ inches } (10 \text{ cm}) \div 20 \text{ inches } (50 \text{ cm})) = 11.45°$$

4. Using the method in Appendix 1, determine the random error of measurement. Record the error in column 10 of your data table.

Results

The angular separation will vary with door width. The author's measurement was 11.45°.

Why?

Like a circle, the sphere of sky surrounding Earth measures 360°. But only about half the sphere is visible above the **horizon** (imaginary line where the sky appears to touch Earth). So the sky you see covers an **arc** (segment of a circle) of about 180° from one side of the horizon to the other. As an observer from Earth, you look at celestial bodies from a great distance. The **apparent distance** between celestial bodies is how large the linear measurement between the bodies appears to be from Earth. The **angular separation** or **angular distance** is the apparent distance expressed in radians or degrees. **Altitude** is the angular distance above the horizon.

The **cross-staff** is an instrument used to determine angular separation. At a specific distance from an object, you measure the apparent width of an object (the width of the sight of the crosspiece) and the distance of the crosspiece from your eye. The ratio of apparent width to distance multiplied by 57.3° expresses the angular separation in degrees.

Try New Approaches

1a. Does the width of the cross-staff's sight affect the results? Repeat the experiment twice, using the medium, "M," and small, "S," sights. Compare your calculations.

b. Assess the accuracy of your cross-staff by determining the true angular separation between the sides of the door (D_A). Do this by using a ruler to measure the actual width of the door, d_a, and using the sighting distance, d_s, of 4 yards (4 m) and this equation:

$$D_A = 57.3° \times (d_a \div ds)$$

For example, if door measurement d_a equals 29 inches, then

$$D_A = 57.3° \times (29 \text{ inches} \div 144 \text{ inches})$$
$$= 11.54°$$

c. Determine the relative error of your measurements using the method in Appendix 2.

2a. Does sighting distance (d_s) affect angular separation? Repeat the original experiment, collecting data for distances of 2, 4, 6, and 8 yards (m) using the wide sight. Determine the relative error for each sighting distance measurement.

b. Repeat the previous experiment using the medium and small sights of the cross-staff. Draw a line graph of your findings. Put the sighting distance on the horizontal axis. Put angular separation on the vertical axis. Use points and lines of different colors for the wide, medium, and small sights.

Design Your Own Experiment

1. Design and experiment to test the accuracy of the cross-staff in measuring angular separation between stars in a **constellation** (group of stars that appears to make a pattern in the sky). For example, you might measure the angular separation of some of the stars in the Big Dipper which is an **asterism** (group of stars with a shape

within a constellation) in the Ursa Major constellation. See separation A shown in Figure 4.2 from Alkaid to Dubhe, separation B from Megrez to Dubhe, and separation C from Dubhe to Polaris. Record your measurements in a table like Table 4.2. If it is too dark to read your cross-staff, stand with the light from a building behind you. Make five measurements for each separation and average them. Compare your average with known (true) angular separations given and record the difference. If the difference is more than the known value, record a positive (+) error. If the difference is less, record a negative (–) error. For more information about the angles between the stars of the Big Dipper, see Terence Dickinson's, *Night Watch* (Willowdale, Ontario: Firefly Books, 1998), p. 30.

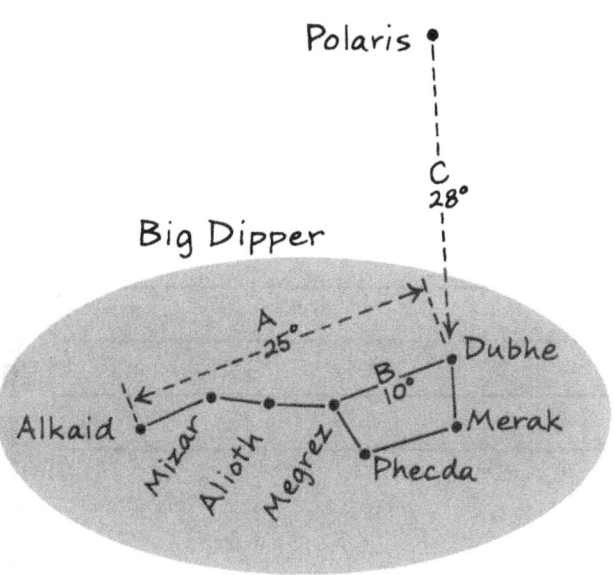

Figure 4.2

2. Design an experiment to determine if the angular separation of celestial bodies changes. Perhaps you can repeat the measurements between the stars of the Big Dipper at different times in one night or at the same time on different nights.

3. Use your cross-staff to measure the angular diameter of a celestial body such as the Moon. Since the Moon's diameter appears to be relatively small, use the smaller notch, S_2, on the left side of the crosspiece. If it is not a full moon, measure the Moon's height rather than its width.

Get the Facts

The cross-staff was the first practical instrument for measuring the altitude of the Sun from the deck of a ship. The cross-staff measured the distance between the horizon and the Sun quite well, but it had problems. The user had to look directly at the Sun, risking eye damage or blindness. Another problem was aligning the crosspiece so that one end appeared to touch the Sun while the other touched the horizon. In 1594, John Davis (1550–1605), a British navigator, published his book *The Seaman's Secrets,* in which he described an improved instrument called the backstaff. What advantages did the backstaff have over the cross-staff? How were these two instruments different? For information about these and other navigational instruments, see Richard Moeschl's *Exploring the Sky* (Chicago: Chicago Review Press, 1993), pp. 115–123. Another resource is Dennis Fisher's *Latitude Hooks and Azmuth Rings: How to Build and Use 18 Traditional Navigational Instruments* (New York: TAB/McGraw-Hill, 1994).

Location	Measured Angular Separation,°						Known Angular Separation, °	Error
	Trial 1	Trial 2	Trial 3	Trial 4	Trial 5	Average		
A							25°	
B							10°	
C							28°	

TABLE 4.2 MEASURING ANGULAR SEPARATIONS WITH A CROSS-STAFF

Notes on
Angular Separation

Key Facts:

_____
_____
_____
_____

My Results:

_____
_____
_____
_____
_____
_____
_____

Conclusions:

_____
_____
_____
_____
_____

Results of Try New Approaches:

_____
_____
_____
_____
_____

Notes on Designing My Own Experiment:

_____
_____
_____
_____
_____
_____
_____
_____
_____
_____
_____
_____
_____
_____
_____
_____
_____
_____
_____
_____
_____

Notes on Get the Facts:

_____
_____
_____
_____
_____
_____
_____

5 Orbital Period
Time of Revolution

German astronomer Johannes Kepler (1571–1630) in his third law of planetary motion expressed the relationship between the orbital period of a planet and its average distance from the Sun. An orbital period is the time it takes a planet to make one revolution, which is once around its orbit.

In this project, you will discover how mass and distance affect a celestial body's orbital period. You will model the effect of the barycenter on the period of a planet. You will also discover how Bode's law predicts the distances of planets.

Getting Started

Purpose: To determine the effect of distance on the orbital period of an orbiting planet.

Materials

⅜-inch (0.93-cm) metal washer	2 pairs of safety goggles
6-foot (1.8-m) cord	timer
ruler	helper

Procedure

Note: This activity is to be performed outdoors.

1. Tie the washer to the end of the cord.

2. On the cord, measure 18 inches (45 cm) from the washer and tie a knot in the cord.

3. Measure 18 inches (45 cm) from the knot and tie a second knot in the cord.

4. Repeat step 3, making a third knot.

5. Put on a pair of safety goggles. In an area away from other people, hold the cord at the first knot—18 inches (45 cm)—from the washer and swing your arm so that the washer spins above your head.

6. Find the slowest speed that will keep the washer "in orbit."

7. Ask your helper to wear safety goggles and be your timekeeper (see Figure 5.1). When your timekeeper says "Start," count the number of revolutions the washer makes. A **revolution** is one turn around an orbit.

Figure 5.1

8. Stop counting when the timekeeper says "Stop," at the end of 10 seconds.

9. Calculate the **orbital period**, T, (time per revolution) of the washer by dividing the time by the number of revolutions. For example, if you counted five revolutions in 10 seconds, the orbital period would be:

$$T = \text{orbital period} = \text{time} \div \text{number of revolutions}$$
$$= 10 \text{ seconds} \div 5 \text{ revolutions}$$
$$= 2 \text{ seconds/revolution}$$

This is read as 2 seconds per revolution and means that it took 2 seconds for the washer to travel 1 revolution.

10. Repeat steps 5 to 9 four times for a total of five trial measurements.

11. Repeat steps 5 to 10 for the two other distances: 36 inches (90 cm) at the second knot, 54 inches (135 cm) at the third knot.

12. Record the data in an Orbital Period versus Orbit Distance table like Table 5.1.

Orbital Distance inches (cm)	Orbital Period (T), seconds/revolution					
	Trial 1	Trial 2	Trial 3	Trial 4	Trial 5	Average
18 (45)						
36 (90)						
54 (135)						

TABLE 5.1 ORBITAL PERIOD VS. ORBIT DISTANCE

13. Make a bar graph of the average orbital periods. Place distance (the independent variable that you changed) on the horizontal axis. Place the orbital period (the dependent variable that changes in response to the independent variable) on the vertical axis. For information about constructing bar graphs, see p. 18.

Results

The longer the cord, the greater the orbital period.

Why?

Kepler's **third law of planetary motion** states that the more distant a planet's orbit from the Sun, the greater the planet's orbital period. This experiment shows that the law works for any object spinning in a circular path.

Try New Approaches

Matter is the substance from which all objects are made. **Mass** is the measure of the amount of matter in an object. How does the mass of the orbiting object affect its orbital period? Repeat the experiment using two washers. As before, provide just enough energy to keep the washers in a circular orbit. In Chapter 6, "Artificial Satellites," page 57, find the orbital **velocity** (speed in a specific direction) of a satellite. Note that **orbital velocity** (revolutions/sec) is the reciprocal of the satellite's orbital period (sec/revolutions). Is the mass of the satellite part of the orbital velocity formula? Do your experimental results agree with the formula?

Design Your Own Experiment

Center of mass is the point on a body where its mass seems to be concentrated. **Gravity** is a force of attraction between all objects in the **universe** (everything throughout space including Earth). Two celestial bodies held together by mutual gravitational attraction are called **binary bodies,** such as the Sun and a planet. The **barycenter** is the point between binary bodies where their center of mass is located, which is the point around which the bodies rotate.

1. The farther the barycenter of a planet-Sun system is from the Sun, the greater the orbital period of the planet. Use two dowels of different lengths to illustrate this principle. First, determine the center of mass of each dowel by balancing it horizontally on your finger. Where your supporting finger touches the rods is their center of mass. Mark the center of mass on each dowel. Rest one end of each dowel on a level surface, such as a floor or table, with a few inches (cm) between them. The flat surface represents the Sun. Hold the dowels up vertically, steadying them with the tips of your fingers. Lean both dowels forward the same amount so they will fall in the same direction. Release both at the same time. Which hits the surface first? For your display you may wish to make a drawing like Figure 5.2 showing the results of this experiment.

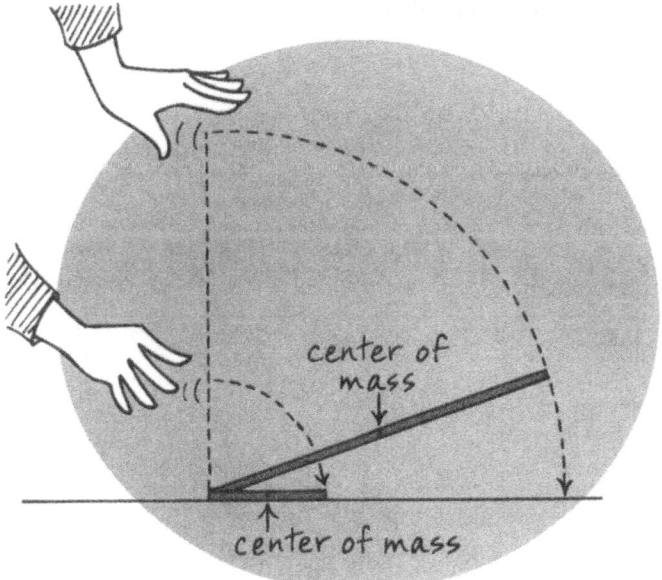

Figure 5.2

2a. Design an experiment to measure the orbital periods of one of the visible planets. Note that Mercury is too close to the Sun to be seen and the movements of Jupiter and Saturn are very slow. The visibility and relatively fast movements of Venus and Mars make them the best subjects. One experiment is to measure the average angular motion of the selected planet per day. Use this to calculate the orbital period, which is the time for 1 revolution of 360°. Staring on day 1, use a cross-staff (see Chapter 4, "Angular Separation," for instructions on making and using a cross-staff) to measure the angular separation between the planet and a star. Take five measurements on the first day and average them. Take five measurements 7 days later and average them. Record measurements in an Angular Separation Data table like Table 5.2. Calculate the orbital period of the planet using these steps:

- Calculate the difference (D_1) between the average angular separation on the first and seventh day.

- Calculate the time (t_1) for D_1, which is the time between the two sets of measurements. Thus, t_1, = 7 days.

- Calculate the portion of the revolutions (D_2) the planet moved in 6 days by dividing D_1 by the degrees in 1 revolution, which is 360°. $D_2 = D_1/360°$.

- Calculate the orbital period (T) using the following equation:

$$T = t_1/D_2$$

TABLE 5.2 ANGULAR SEPARATION DATA

	Angular Separation, °					
	Trial 1	Trial 2	Trial 3	Trial 4	Trial 5	Average
D_1						
D_2						

b. Kepler's third law of planetary motion states that a planet's orbital period (time of 1 revolution around the Sun) depends on its average distance from the Sun. This equation is:

$$P^2 = k \times r^3$$

where P is the orbital period, r the average distance from the Sun, and k a constant. If P is in years and r is in AU, k equals 1. Using the known radius of the visible planets (see Appendix 3), calculate the orbital period of each. Note that the radius (½ diameter) must be expressed in astronomical units (AU), which can be determined by taking the average distance of the planet from the Sun and dividing by the average distance of Earth from the Sun. For example, for Mercury the radius in AU is 0.39 AU. See page 33 for AU calculations.

c. Using the known orbital period and the experimentally determined orbital periods in the previous experiment, calculate your experimental percentage error. (See Appendix 2 for information about percentage error.) What is retrograde motion and how would it affect your percentage error? For information about retrograde motion, see Dinah Moche's *Astronomy* (New York: Wiley, 2000), pp. 197–200.

Get the Facts

The German astronomer Johann Titius (1729–1796) showed that the distances of planets from the Sun follow a fixed formula when measured in astronomical units. The formula is known as *Bode's law.* What is this pattern? Why isn't it called Titius's law? What significant role did the pattern play in the discovery of the asteroids and some of the planets? How accurate are the distances using the formula? For information, see Nancy Hathaway's, *The Friendly Guide to the Universe* (New York: Penguin, 1994), pp. 190–192.

5

Notes on
Orbital Period

Key Facts:

_____
_____
_____
_____

My Results:

_____
_____
_____
_____
_____
_____
_____
_____

Conclusions:

_____
_____
_____
_____
_____

Results of Try New Approaches:

_____
_____
_____
_____
_____

Notes on Designing My Own Experiment:

_____
_____
_____
_____
_____
_____
_____
_____
_____
_____
_____
_____
_____
_____
_____
_____
_____
_____
_____
_____
_____

Notes on Get the Facts:

_____
_____
_____
_____
_____
_____

6 Artificial Satellites
Man-Made Orbiters

A satellite is a celestial body that orbits another. An artificial satellite is a man-made object that orbits Earth. Artificial satellites are raised to a desired height above Earth and launched by rockets parallel to Earth's surface. This forward velocity (speed of an object in a specific direction) and the force of gravity keep the satellite in orbit around Earth.

In this project, you will learn about launching speed and its effect on a man-made satellite's orbit. You will discover the best times to see satellites and how to measure their angular velocity. You will also find out why satellites are launched in different directions.

Getting Started

Purpose: To model how an artificial satellite is launched into orbit.

Materials

2 equal-size books, each about 10 inches (25 cm) long
index card
transparent tape
3 rulers—2 must be identical and have a groove down the center

walnut-size piece of modeling clay
bath towel
marble

Procedure

1. Lay the books end to end on a table.

2. Lay the index card on the end of the books farther from the edge of the table.

3. Tape the grooved rulers together end to end. Tape only on the ungrooved side. This is your launcher.

4. Lay the launcher on the books so that one end is on the index card. Raise that end 2 inches (5 cm) above the book. Put the clay under the end for support. Let the other end of the launcher extend over the end of the books.

5. Adjust the books so that the extended end of the launcher is 4 inches (10 cm) from the edge of the table.

6. Lay the towel on the floor near the edge of the table. The towel will help stop the marble when it hits the floor.

7. Position the marble on the raised end of the launcher, then release it (see Figure 6.1) The marble should land on the towel.

8. Observe the path of the marble after it leaves the launcher.

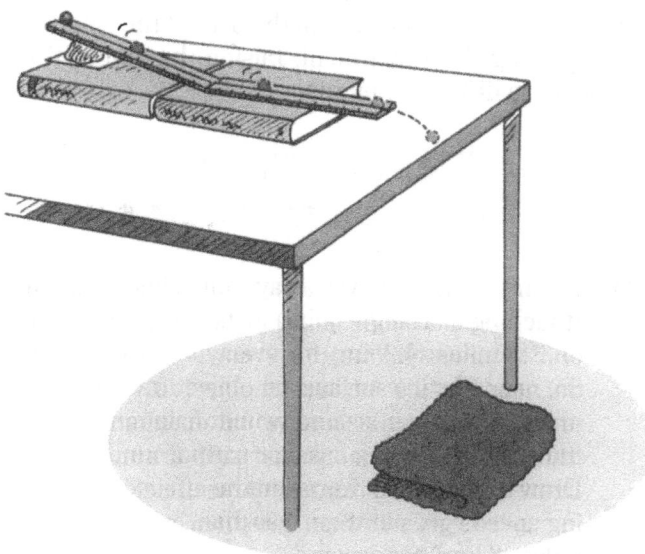

Figure 6.1

Results

The marble's path curves after it leaves the launcher.

Why?

An **artificial satellite** (man-made satellite) is any object purposely placed into orbit around a celestial body, including Earth. Since the launching of the first artificial satellite in 1957, thousands of artificial satellites have been placed in orbit around Earth. These satellites have many uses for private companies, governments, and educational facilities. This experiment models the launching of an artificial satellite, called the marble satellite.

In this model, the table represents Earth. The top of the books is a position above Earth's surface where the "marble satellite" is launched horizontally, parallel to Earth's surface. After it separates from its launcher, the satellite moves in a curved path. The curve results from the satellite's forward horizontal velocity and the downward pull of gravity. A real satellite would continue revolving around Earth in a curved path called an orbit that would pass through the launching point.

Try New Approaches

If the horizontal velocity of the marble satellite is great enough, gravity pulls it into a curved path past the edge of the table. Demonstrate the effect of different horizontal velocities on the path of the satellite by repeating the experiment, raising the end of the launcher to different heights.

Design Your Own Experiment

1a. Earth's surface curves away from a line **tangent** (touching at a single point) to its surface at a rate of $3\frac{3}{50}$ miles (4.9 km) for every 5 miles (8 km). So, near Earth's surface, an object traveling at 5 miles (8 km) per second would maintain its altitude and move in a circular path around Earth. Draw a diagram to represent the effect of launching speeds greater than, less than, and equal to 5 miles (8 km) per second.

b. A satellite's horizontal speed depends on its distance above Earth, which affects the strength of gravity acting on the satellite. The gravity at a certain distance from the center of Earth can be calculated by using this equation:

$$g_1/g_2 = r_2^2/r_1^2$$

where g_1 = 9.8 m/s^2, gravity at distance equal to the radius of earth (at surface)

r_1 = 12,757 km, radius of Earth

g_2 = gravity at distance r_2

r_2 = distance from center of Earth to satellite's orbit

The least velocity a satellite must have to orbit Earth is determined by this equation:

$$V = \sqrt{gr}$$

where V = the velocity of the satellite

g = the acceleration of gravity at distance r from the center of Earth

r = the average radius of the satellite's orbit from the center of Earth

Artificial satellites that stay above one place on Earth as they orbit are called **geosynchronous satellites**. These satellites orbit above Earth at an **altitude** (height above a surface) of 22,300 miles (13,938 km). Use these formulas to determine the velocity of a geosynchronous satellite and the gravity acting on it. **Science Fair Hint:** Prepare a diagram showing satellites at different distances from Earth's surfaces, the calculations used to determine their velocity, and the gravity (force of attraction between all objects in the universe) acting on them. For more information, look up satellite velocity in a physics text.

2. Satellites look like slow-moving stars. Those in orbits at an altitude of 300 miles (480 km) or less are visible to the naked eye. Determine when satellites are most visible. Observe satellites at different times of the night and during different seasons. For more information on satellite motion, see Terence Dickinson's *Night Watch* (Willowdale, Ontario: Firefly Books, 1999), p. 35.

3. Determine the **angular velocity** (angular displacement per unit time of an object moving in a curved path) of satellites by measuring the angular distance traveled during a measured amount of time. Find out which satellites are visible on a certain date from sources such as *Sky and Telescope* magazine or from Web sites such as www.skyandtelescope.com. (See Part II, Chapter 4, "Angular Separation," for information about measuring angular distance.)

Get the Facts

Most satellites are launched from west to east, but some are launched to orbit Earth from pole to pole. Why is a launch into a polar orbit more difficult? What are satellites used for? For information, see Dinah Moche's *Astronomy Today* (New York: Random House, 1995), pp. 20–21.

Notes on
Artificial Satellites

Key Facts:

_____

_____

_____

_____

My Results:

_____

_____

_____

_____

_____

_____

_____

Conclusions:

_____

_____

_____

_____

_____

Results of Try New Approaches:

_____

_____

_____

_____

_____

Notes on Designing My Own Experiment:

_____

_____

_____

_____

_____

_____

_____

_____

_____

_____

_____

_____

_____

_____

_____

_____

_____

_____

_____

_____

_____

_____

_____

_____

Notes on Get the Facts:

_____

_____

_____

_____

_____

7 Rotation
The Spinning of Celestial Bodies

Rotating celestial bodies spin around an axis, an imaginary line running through their centers. In 1851, French physicist Jean-Bernard-Léon Foucault (1819–1868) interpreted the motion of a pendulum as proof that Earth rotates on its axis. A pin at the end of his pendulum made marks in sand on the floor of the Panthéon in Paris. As the minutes passed, the direction of the pendulum remained the same, but the marks underneath it changed. This proved that the floor moved beneath the pendulum as a result of Earth's rotation.

In this project, you will determine how a pendulum moves at the North and South Poles. You will determine if the length of a Foucault pendulum affects the results. You will also learn how the apparent shift in the path of a pendulum varies at different locations on Earth.

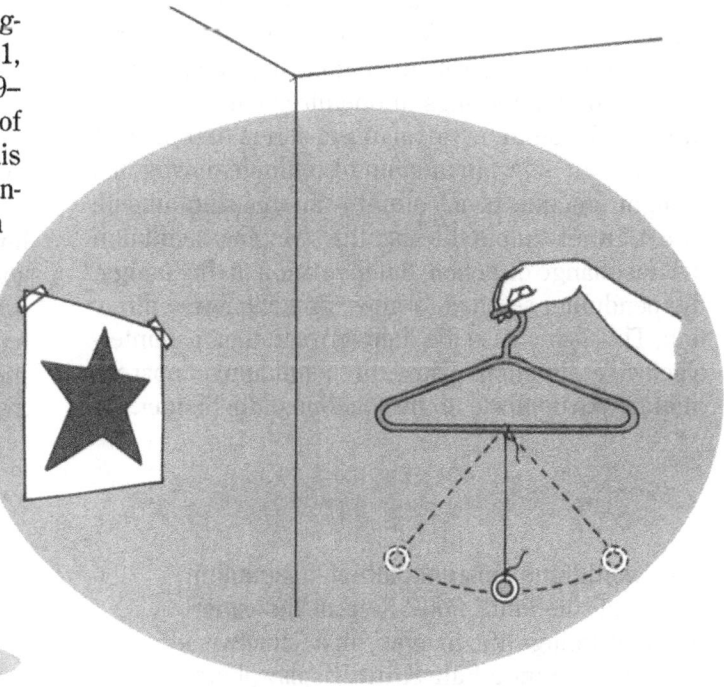

Figure 7.1

Getting Started

Purpose: To model the motion of a pendulum at the North Pole.

Materials

marker	10-inch (25-cm) string
sheet of white copy paper	metal washer
masking tape	clothes hanger

Procedure

1. Draw a large star on the paper.

2. Tape the paper to a wall.

3. Tie one end of the string to the washer.

4. Tie the free end of the string to the center of the hanger.

5. Holding the hanger so that one end of it faces the star, lift the washer in the direction opposite the star on the paper. Release the string and let it swing toward and away from the star (see Figure 7.1).

6. While the washer is swinging, slowly rotate the hanger about one-fourth of a turn in a counterclockwise direction. Observe the direction that the swinging string moves in relation to the hanger.

7. Rotate the hanger another one-fourth of a turn in a counterclockwise direction and again observe the direction that the swinging string moves in relation to the hanger.

8. Repeat step 7 two times, bringing the hanger back to its original position.

Results

The string continues to swing toward and away from the star. In relation to the hanger, the pendulum appears to move in a clockwise direction.

Why?

This investigation represents the motion of a **pendulum** (a weight that is suspended from a point and is free to swing back and forth) placed at the North

Pole. The imaginary north-to-south line through the center of a body and about which it rotates is called an axis. The ends of the axis of a celestial body are called **poles.** The north and south end of an axis are called the North Pole and South Pole, respectively. The clothes hanger represents Earth rotating (turning on its axis). As viewed from above the North Pole, Earth rotates in a counterclockwise direction, like the turning hanger. When set in motion, the washer-and-string pendulum swings in one direction because of **inertia** (the tendency of an object at rest to remain at rest or of an object in motion to continue moving in a straight line unless acted on by an opposing outside force). In relation to the star, the swinging pendulum did not change direction, but in relation to the hanger, the pendulum seemed to move in a clockwise direction. This is because the hanger rotated in a counterclockwise direction. Thus, the pendulum's apparent motion was opposite to the rotation of the hanger.

Try New Approaches

How would the apparent path of a pendulum appear at the South Pole? Repeat the experiment, rotating the hanger in a clockwise direction. **Science Fair Hint:** Display photographs with labels indicating clockwise and counterclockwise rotations to represent the pendulum's swing at the poles.

Design Your Own Experiment

1a. Foucault's pendulum was a 220-foot (67-m) wire holding a 62-pound (28-kg) sphere. Find out if a shorter, lighter pendulum would work as well. Suspend a wire from a tree limb. To reduce friction and rotation, use two pieces of wire, one short, the other long. Tie both wires to a fishing swivel. Attach the short section of the wire to the limb and the longer section to a weight, such as a heavy bucket. Place a piece of poster board with a line drawn across it beneath the weight (see Figure 7.2). Set the pendulum in motion in the direction of the line. Observe the motion of the pendulum in relation to the line on the poster board.

b. Determine if the length of the wire affects the outcome. Use varied lengths of wire with the same weight.

c. Find out if the weight makes a difference. Keep the lengths of wire the same, but attach heavier or lighter objects.

Get the Facts

The path traced by Foucault's pendulum in Paris, at latitude 48°N, appeared to shift more than 10° per hour. (*Latitude* is angular distance in degrees north and south of the equator.) The degree of shift depends on latitude. Find out why. Learn how to calculate the shift for any given latitude. For information, see *Janice VanCleave's A+ Projects in Earth Science* (New York: Wiley, 1999), pp. 27–28.

Figure 7.2

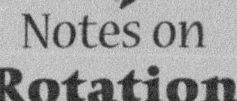

Notes on
Rotation

Key Facts:

_____

_____

_____

_____

My Results:

_____

_____

_____

_____

_____

_____

_____

Conclusions:

_____

_____

_____

_____

Results of Try New Approaches:

_____

_____

_____

_____

_____

Notes on Designing My Own Experiment:

_____
_____
_____
_____
_____
_____
_____
_____
_____
_____
_____
_____
_____
_____
_____
_____
_____
_____
_____
_____
_____
_____
_____
_____
_____
_____
_____

Notes on Get the Facts:

_____
_____
_____
_____
_____
_____

8 Apparent Magnitude
Apparent Star Brightness

Stars vary widely in brightness. Some appear very bright, while others are barely visible to the naked eye. Around 150 B.C., long before the invention of telescopes, the Greek astronomer Hipparchus devised a scale to measure apparent magnitude, the brightness of stars as seen with the naked eye from Earth. He gave a value of 1 to the brightest star and a value of 6 to the dimmest. Today, we use a variation of his scale to measure the brightness of stars. Instead of observing and estimating magnitudes with the naked eye, we now use an instrument called a photometer, which produces more precise measurements. Also, the scale has been extended beyond 1 to 6 so astronomers can measure an even broader range of brightness.

In this project, you will demonstrate the effect of luminosity and distance on the apparent magnitude of a star. You will build an instrument to measure apparent magnitude. You will learn how apparent magnitude differs from intrinsic (natural) luminosity, which is the amount of light a star emits. You will also discover the difference between apparent and absolute magnitude.

Getting Started

Purpose: To demonstrate how distance affects the brightness of an object.

Materials

3 pencils
yardstick (meterstick)

2 identical incandescent flashlights with new batteries
2 helpers

Procedure

1. In an open area outdoors, stick a pencil in the ground to mark the starting point. Use the yardstick (meterstick) to measure two distances from the pencil, one at 10 feet (3 m) and the second at 30 feet (9 m). Mark these distances with pencils in the ground.

2. At night, stand beside the first pencil.

3. Ask your helpers to hold flashlights and to stand side by side at the second pencil, 10 feet (3 m) away.

4. Instruct your helpers to turn on their flashlights and shine them toward you.

5. Look at the lights just long enough to compare their brightness.

6. Ask one of your helpers to move to the third pencil, 30 feet (9 m) away, while continuing to shine the light toward you (see Figure 8.1).

7. Again, compare the brightness of the lights.

8. Ask your other helper to move to the third pencil while continuing to shine the light toward you.

9. As before, compare the brightness of the lights.

Figure 8.1

Results

The lights appear equally bright at an equal distance from you. When they are at different distances, the closer light appears brighter.

Why?

Magnitude is a measure of how bright a celestial body appears to be. **Apparent magnitude** is a measure of how bright a celestial body appears as viewed with the naked eye from Earth. Apparent magnitudes are ranked on a **magnitude scale,** with an inverse relationship between brightness and magnitude numbers, expressed as magnitudes. For example, the magnitude 1 star in Figure 8.2 is brighter than the magnitude 3 star. Apparent magnitude is not a measure of luminosity. **Luminosity** is the amount of light energy a light source such as a star gives off in a given amount of time. When stars have the same luminosity, the closer star, like the closer flashlight, has a greater magnitude.

Try New Approaches

The star Aldebaran, in the constellation Taurus, is about 50 times larger and about 100 times more luminous than the Sun. It appears far less bright, however, because it is so far away from Earth. Demonstrate how distance affects the apparent magnitude of stars with unequal luminosity. Repeat the experiment using two flashlights, fresh batteries in one and weak batteries in the other.

Design Your Own Experiment

1. Design a model to illustrate how distance affects the apparent magnitude of stars with equal luminosity. One way is to make a drawing using lines to indicate luminosity. In Figure 8.2, the three stars emit the same amount of energy, as indicated by the number of lines (6) in the boxes. The numbers show, however, that their apparent magnitude varies, with the closest star appearing the brightest (having the lowest lowest magnitude number).

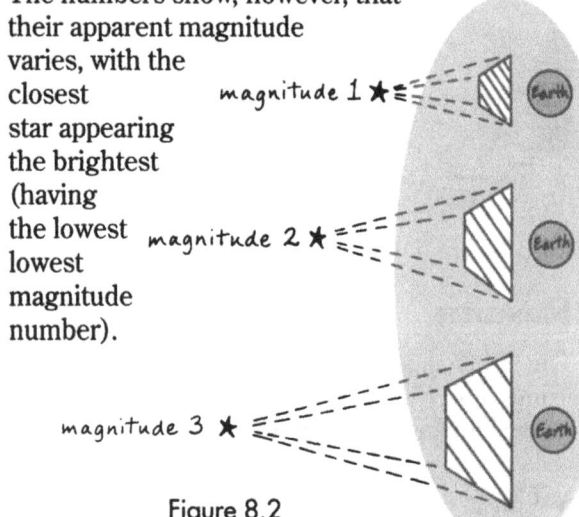

Figure 8.2

2a. Design an experiment to measure the apparent magnitude of stars. One way is to make a brightness viewer. The following steps provide a method for testing the brightness of a magnitude 1 star.

- Punch two holes near the short edge of an index card. Label the card "Magnitude 1."

- Cover one of the holes with a piece of transparent (not frosted) tape (see Figure 8.3). The covered hole is the testing hole. The uncovered hole is the viewing hole.

Figure 8.3

- To calibrate the viewer:

  (1) Through the viewing hole, look at a star known to be of magnitude 1. For information about star magnitude, see Terence Dickinson's *Night Watch* (Willowdale, Ontario: Firefly Books, 1998), pp. 100–119.

  (2) Look at the same star through the testing hole. The star should be barely visible. If it appears bright, add more layers of tape until the star is barely visible. ***Note:*** Be careful to keep the layers of tape clean and wrinkle free.

- To use the viewer, find a star by looking through the viewing hole. Then move the viewer so that you look at the same star through the testing hole. If you can barely see the star, it has a magnitude of 1 or less.

b. To identify stars with magnitudes greater than 1, repeat the previous experiment for stars of greater magnitude.

3. The stars that make up the bowl of the constellation Ursa Minor, commonly called the Little Dipper, have magnitudes of 2, 3, 4, and 5 (see Figure 8.4). Use these stars as a "magnitude scale in the sky" against which the magnitudes of other stars can be compared and estimated.

Figure 8.4

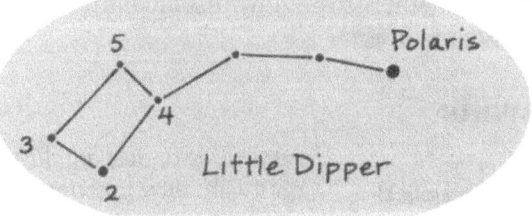

Get the Facts

Some stars give off much more light than the Sun, but Earth receives thousands of millions of times more light from the Sun than from any other star. Why? To compare the light outputs of the Sun and other stars, astronomers use an *absolute magnitude* scale. How was it developed? How does it work? For information, see the *National Audubon Society Field Guide to the Night Sky* (New York: Knopf, 1995), pp. 24–25.

Notes on
Apparent Magnitude

Key Facts:

_____
_____
_____
_____

My Results:

_____
_____
_____
_____
_____
_____
_____
_____

Conclusions:

_____
_____
_____
_____
_____

Results of Try New Approaches:

_____
_____
_____
_____
_____

Notes on Designing My Own Experiment:

_____
_____
_____
_____
_____
_____
_____
_____
_____
_____
_____
_____
_____
_____
_____
_____
_____
_____
_____
_____
_____
_____

Notes on Get the Facts:

_____
_____
_____
_____
_____
_____

9 Parallax
Apparent Shift of an Object

When an observer changes viewing positions, a nearer object appears to move somewhat in relation to objects in the distant background. Scientists call this apparent movement parallax. Astronomers use parallax to measure the distance of stars.

In this project, you will discover how two factors, the distance from an object and the baseline (distance between observing points), affect parallax. You will learn how to measure the distance to a nearby object using parallax shift. You will also find out how to measure the distance to faraway celestial bodies.

Getting Started

Purpose: To determine how the distance to an object affects parallax.

Materials

marking pen
yardstick (meterstick)
22-by-28-inch (55-by-70-cm)
   sheet of white poster board
masking tape
7-foot (2.1-m) string
flat toothpick

Procedure

1. Draw 27 lines 1 inch (2.5 cm) apart across the poster board parallel to the short ends. Number the lines from left to right.

2. Tape the poster board to the edge of a table so that the lines run vertically.

3. Measure 14 inches (35 cm) from one end of the string. Place a piece of tape on the string at this point. Label the piece of tape "1."

4. From the tape on the string, measure 1 foot (30 cm) and place a second piece of tape. Label the piece of tape "2." Repeat this step four times, numbering the pieces of tape in order. There should be six pieces of tape on the string.

5. Place the yardstick (meterstick) so that about 6 inches (15 cm) of it extends over the center of the poster board. Tape the measuring stick to the table.

6. Tape the toothpick to the extended end of the measuring stick, pointing down.

7. Lay 2 inches (5 cm) of the end of the string that is near tape 1 on the extended end of the stick. Tape the string to the stick.

8. Sit on the floor in front of the toothpick. Stretch the string and adjust your sitting position so that label 1 touches your nose. In this position, your eyes are in line with and about 1 foot (0.3 m) from the toothpick.

9. Close your left eye and look at the toothpick with your right eye (see Figure 9.1). Move your head until the toothpick aligns with one of the vertical lines on the poster board. Note the number of that line.

Figure 9.1

10. Without moving your head, open your left eye and close your right eye. Note the number of the line the toothpick appears to move to. If the toothpick falls between lines, estimate the line number to the nearest tenth.

11. Record the parallax measurement in a table like Table 9.1. Determine the parallax by finding the difference between the positions of the toothpick as seen with your right and left eyes. For example, if you see the toothpick in front of line 4 with your right eye and in front of line 9 with your left eye, the parallax equals 9 minus 4, or 5 spaces.

12. Repeat steps 8 to 11 at each of the marked distances on the string.

TABLE 9.1 PARALLAX EFFECT			
Distance from the Object, feet (m)	Right Eye Position	Left Eye Position	Parallax, spaces
1 (0.3)	9	4	5
2 (0.6)			
3 (0.9)			
4 (1.2)			
5 (1.5)			
6 (1.8)			

Results

Parallax decreases as distance from the object increases.

Why?

Each eye sees the toothpick from a different point. The distance between the points is the **baseline.** The apparent change in position of an object when viewed from two different points is called **parallax.** As the distance from the toothpick increased, parallax decreased. The farther away an object, the smaller the measurement of parallax.

Try New Approaches

How does the length of the baseline affect parallax? Look in a mirror and measure the distance between the pupils of your eyes. This was the baseline for the original experiment. Make a viewer with baselines greater and less than your eye baseline. Draw a 5-inch (12.5-cm) line across the widest part of a 4-by-6-inch (10-by-15-cm) index card. Punch a hole at each end of the line. Label each hole "1." In the center of the card, punch two holes 1 inch (2.5 cm) apart and label each hole "2." Repeat the experiment twice, first looking through the number 1 holes, then through the number 2 holes. To look through the holes, begin by holding the card so that its center is in line with the end of your nose. Keeping the card in place, move your head to the right and left to look through the holes.

Design Your Own Experiment

1. Design an experiment to demonstrate how parallax effect is used to measure the distance (d) to a nearby object. One way is to use a predetermined baseline and measure the parallax shift. Then determine the distance to the object using the following equation:

$$d = 57.3° \text{ (baseline distance} \div \text{ parallax shift)}$$

Note: This equation yields a number without a unit of measurement. When no unit of measurement is indicated in giving the measure of an angle, the angle is understood to be expressed in radians. To express the angle in degrees, the conversion 57.3° per 1 radian is used.

Mark two points on the ground exactly 10 feet (3 m) apart and about 30 paces from a tree. Label the points "A" and "B." Stand at point A and use a cross-staff to measure the angular separation between the tree and a distant object such as a telephone pole. (See Chapter 4 for more information about angular separation and using a cross-staff to measure it.) ***Note:*** The distant object (telephone pole) should be about 10 times or more distant than the near object (tree) whose distance is being measured. Move to point B and make another measurement (see Figure 9.2). For the example shown, if the angular separation measures 5° from point A and 5° from point B, then the parallax shift equals their sum, or 10°. The distance to the tree is then calculated as follows:

$$d = 57.3° \text{ (10 feet} \div 10°) \text{ or } 57.3° \text{ (3 m} \div 10°)$$
$$= 57.3 \text{ feet (17.2 m)}$$

b. Use a tape measure to measure the distance to the tree. Then use the method in Appendix 2 to

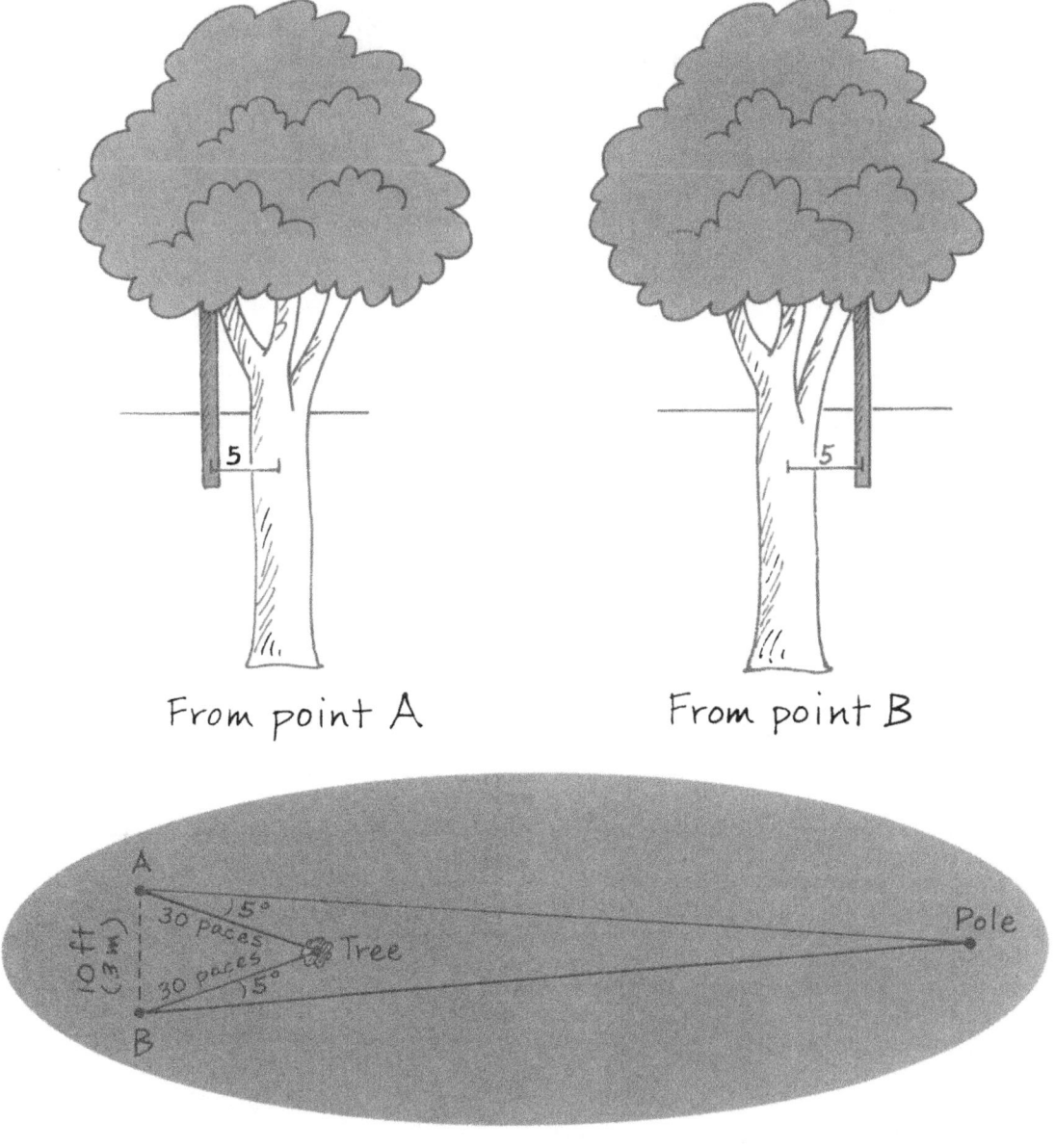

From point A   From point B

Figure 9.2

determine the relative error of your measurements. This will give you an indication of how accurate your cross-staff is.

2. Diagram the parallax shift of a star using the diameter of Earth's orbit as the baseline. For information as well as other parallax experiments, see *Janice VanCleave's A+ Projects in Earth Science* (New York: Wiley, 1999), pp. 43–50.

Get the Facts

The parallax method cannot measure the distance of most stars because they are too far away. Instead, astronomers use *photometers* (light meters), *Cepheids* (pulsating stars), and, for stars in remote galaxies, something called red shift. For information on how these methods are used to calculate distances to stars, see *Terence Dickinson, Nightwatch: A Practical Guide to Viewing the Universe* (Willowdale, Ontario: Firefly Books, 1998), p. 89.

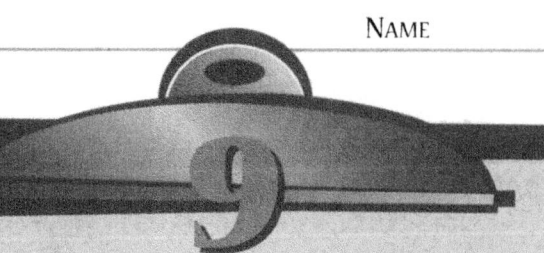

Notes on
Parallax

Key Facts:

_____
_____
_____
_____

My Results:

_____
_____
_____
_____
_____
_____
_____
_____

Conclusions:

_____
_____
_____
_____
_____

Results of Try New Approaches:

_____
_____
_____
_____

Notes on Designing My Own Experiment:

_____
_____
_____
_____
_____
_____
_____
_____
_____
_____
_____
_____
_____
_____
_____
_____
_____
_____
_____
_____
_____
_____
_____
_____

Notes on Get the Facts:

_____
_____
_____
_____
_____
_____
_____

10 Sunspots
Cooler Surface Regions

In the early 1600s in Paris, a German astronomer and Jesuit priest, Christoph Scheiner (1573–1650), was convinced he saw dark spots on the Sun. He was criticized for that idea because the renowned Greek philosopher Aristotle (384–322 B.C.) had stated that everything in the universe except Earth was perfect and without flaws. Scheiner revised his opinion and stated that the spots were caused by something near but not on the Sun. However, Galileo (1564–1642), an Italian scientist, saw the sunspots, studied their motion, and argued that they originated on the Sun. He is often credited as the first to discover them.

In this project, you will create a model of the layers of the Sun. You will investigate the photosphere, the area that appears to be the Sun's surface. You will observe the movement of sunspots to confirm that the Sun rotates. You will also learn how sunspots cycle and how they affect the Sun's activity.

Getting Started

Purpose: To build a model of the Sun's internal structure.

Materials

serrated knife (use with adult approval)

Styrofoam ball, 6 inches (15 cm) in diameter or larger

2 different colored permanent markers

four 1-by-4-inch (2.5-by-10-cm) white labels

4 round toothpicks

pen

Procedure

1. Use the knife to cut away one-fourth of the Styrofoam ball. Set the ball aside.

2. With one of the markers, paint an area in the center of the ball to represent the Sun's core.

3. Use the second marker to draw a band surrounding the core to represent the convection zone.

4. Prepare four flags using these steps:

- Touch the two sticky ends of a label together, leaving a gap near the folded end.

- Insert a toothpick through the gap and press the label onto the toothpick to make a flag (see Figure 10.1).

- Write the names of the Sun's layers on the flags: "Photosphere," "Convection Zone," "Radiation Zone," "Core."

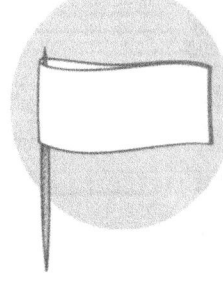

Figure 10.1

5. Stick the flags in the Sun model as shown in Figure 10.2.

6. Make a Layers of the Sun table like Table 10.1, showing the thickness and temperature of each layer.

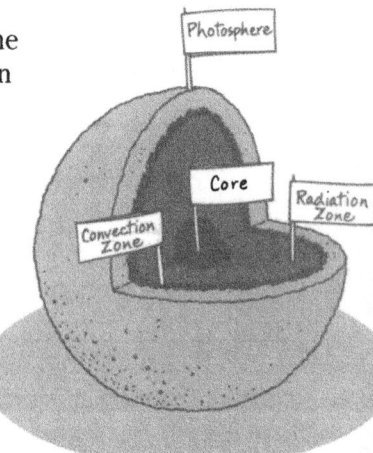

Figure 10.2

Layer	Thickness, miles (km)	Temperature, °F (°C)
Core	87,000 (139,200)	27,000,032 (15,000,000)
Radiation Zone	239,250 (382,800)	4,500,032 (2,500,000)
Convection Zone	108,750 (174,000)	1,980,032 (1,100,000)
Photosphere	342 (547)	9,932 (5,500)

TABLE 10.1 LAYERS OF THE SUN

Results

You have made a model showing the layers of the Sun and a table describing some of their characteristics.

Why?

The Sun's **core** (center of a celestial body) is its hottest part. In the core, **nuclear fusion** (the combing of the nuclei of atoms) releases enormous amounts of **energy** (ability of matter to make changes) called radiation. **Radiation** is energy that can move through space, and is not carried by matter. Energy from the hot core moves through the area outside the core called the **radiation zone.** From there, heated gases expand and rise. When they cool, they become denser and sink back down again. Circulating gas forms the **convection zone.** The next layer, the **photosphere,** is actually the first layer of the Sun's **atmosphere** (gaseous layer surrounding any celestial body that has gravity strong enough to prevent the gases from escaping). But from Earth, it looks like the Sun's surface. Your model does not show the layers of the atmosphere above the photosphere: the **chromosphere** and the outermost layer, the **corona.**

Try New Approaches

Photographs of the Sun's photosphere show that it has a grainy appearance called **granulation.** Covering the photosphere are bright spots resembling rice grains, called **granules.** A granule is the top of a rising current of hot gases from the convection zone. A single granule measures about 900 miles (nearly 1,500 km) across. Between the granules are dark **intergranular lanes** where the cooler gases return to the convection zone. Also seen on the photosphere are large dark spots called **sunspots.** These are centers of intense localized magnetic fields that are thought to suppress the currents of hot gases. They are dark because they are cooler than the areas surrounding them. Find out more about granules and sunspots and add them to your Sun model. For information about the size of sunspots and their two observable regions, the **umbra** (darkest part) and the **penumbra** (grayish outer part), see Dinah Moche's *Astronomy: A Self-Teaching Guide* (New York: Wiley, 1996), pp. 102–105.

Design Your Own Experiment

The Sun rotates about once every 4 weeks. Design an experiment to confirm that the Sun rotates. Observe and record the location of sunspots over a period of time, but unlike Galileo, who looked directly at the Sun and in time lost his eyesight, *you must NOT look directly at the Sun.* A safe, indirect method of observing

sunspots uses a telescope, but binoculars with one lens covered will work, too. *Without looking through the lens,* point the telescope at the Sun and adjust the position of the instrument until its shadow is as small as possible. Then focus the Sun's image on a sheet of white poster board—the screen. Secure the telescope and screen so that they do not move by setting the telescope on a tripod and tacking the screen to a tree (see Figure 10.3). Cut a hole in a second piece of poster board large enough to fit around the end of the telescope that faces the Sun. This sheet of poster board casts a shadow on the screen, making the Sun's image easier to see. At the same time each day or as many days as possible for 2 weeks or more, tape a clean sheet of paper to the screen. Mark the Sun's image and the location of sunspots on the paper. Compare the drawings. How much do the sunspots move? Do some move faster than others? Does the path of sunspots reveal the location of the Sun's equator and poles? For more information, see Philip Harrington's *Astronomy for All Ages* (Old Saybrook, CT: Globe Pequot Press, 1994), pp. 77–79. **CAUTION:** *Never look directly at the Sun. It will permanently damage your eyes.*

Figure 10.3

Get the Facts

At any given moment, the number of visible sunspots can vary from several hundred to none at all. The number increases and decreases in a regular pattern called the *sunspot cycle.* How long does each cycle last? How does the Sun's activity relate to the number of sunspots? Get today's sunspot number at www.spaceweather.com. For more information, see Moche's *Astronomy,* p. 103.

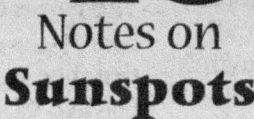

Notes on
Sunspots

Key Facts:

_____

_____

_____

_____

My Results:

_____

_____

_____

_____

_____

_____

_____

_____

Conclusions:

_____

_____

_____

_____

_____

Results of Try New Approaches:

_____

_____

_____

_____

_____

Notes on Designing My Own Experiment:

_____
_____
_____
_____
_____
_____
_____
_____
_____
_____
_____
_____
_____
_____
_____
_____
_____
_____
_____
_____
_____
_____
_____
_____
_____
_____
_____

Notes on Get the Facts:

_____
_____
_____
_____
_____
_____

11 Galilean Satellites
Jupiter's Largest Moons

The four largest satellites of Jupiter are called Galilean satellites because they were discovered in 1610 by Galileo. Their names are Io, Europa, Ganymede, and Callisto, from closest to farthest from Jupiter.

In this project, you will discover the motion of each Galilean satellite during an Earth day (24 hours). You will plot the positions of the satellites as they move around Jupiter and determine their locations relative to Jupiter from a bird's-eye view above Jupiter's north pole, as well as from an observer's view from Earth. From actual observations, you will identify the Galilean satellites by name and learn more about each one.

Getting Started

Purpose: To determine the motion of Io during an Earth day (24 hours).

Materials

sheet of white copy paper   metric ruler
drawing compass   pencil
                                 protractor

Procedure

1. The radius of Io's orbit is about 4.9×10^5 km. Using a scale of 6 mm/1×10^5 km, calculate the radius of the orbit for a scale model as follows:

actual radius of Io's orbit × scale

4.9×10^5 km × 6 mm/1×10^5 km = 29.4 mm

Rounding the number to the closest millimeter, the radius of the scale model of Io would be 29 mm.

2. In the center of the paper, use the compass to draw a circle with a radius of 29 mm. The circle represents the orbit of Io.

3. In the center of the circle, make a small circle with a diameter of about 8 mm. This circle represents Jupiter.

4. Label the directions "East" and

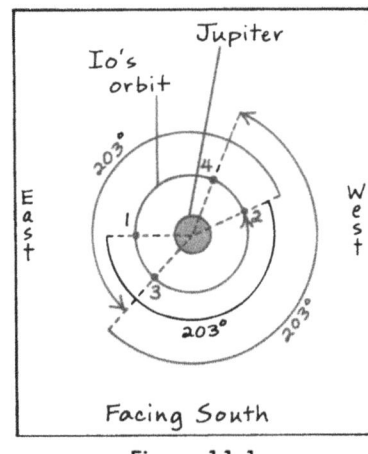

Figure 11.1

"West" on the paper as shown, in Figure 11.1 with "Facing South" at the bottom of the paper.

5. Make a dot on the east side of the large circle. Number the dot 1. This dot represents the first observation location of Io.

6. To find the angular distance of Io's movement in one Earth day (24 hours), divide the angular distance of one revolution (360°) by Io's orbital period (1.77 days). Round the answer to the nearest whole number of degrees.

$$d = 360° \div 1.77 \text{ days} = 203°$$

7. Use the protractor to find a point on the large circle 203° from the first dot. Make a dot on the circle at this point and number the dot 2.

8. Use the protractor to find a point on the circle 203° from dot 2. Make a dot on the circle at this point and number the dot 3. Repeat to find the location of dot 4.

9. Use the compass to draw 203° arcs as shown in Figure 11.1 to indicate the angular distance between the dots.

Results

The dots represent Io's position after 0, 1, 2, and 3 Earth days, with observation beginning at position 1.

Why?

A natural satellite is a celestial body revolving around another larger celestial body, such as the moon Io around Jupiter. The orbital radius of Io is about 4.9×10^5 km (3.1×10^5 miles). In this experiment you made a scale model of Jupiter's moons. A scale is a ratio between the measurements of a model and the actual measurements of an object. In a scale ratio the model measurement comes first. If the first value of the ratio is smaller, the scale drawing or model is reduced, such as the 6 mm/1×10^5 km in this activity. At this scale, the small center circle approximates Jupiter's diameter of about 1.4×10^5 km (0.875×10^5 miles). The distance of 29 mm from the center of the circle approximates the orbital distance of Io from Jupiter's center.

The locations of dots 1, 2, 3, and 4 indicate the movement of Io over 3 Earth days (72 hours). During the first Earth day, Io moved counterclockwise from dot 1 to dot 2, moving 203° around Jupiter. During the second Earth day, Io moved from dot 2 to dot 3, another 203°. Then during the third day, Io moved from dot 3 to dot 4, another 203°.

Try New Approaches

Assuming all of the Galilean moons start at the same point on the first day of observation, how would their apparent distances from Jupiter appear at the end of 2 Earth days? Repeat the project using a data table like Table 11.1 to determine the angular distance each moon moves during 1 Earth day. Draw orbits at the correct radius from Jupiter's core using the same scale (6mm/1×10^5 km). Use different-colored markers to represent each satellite. One at a time, plot the positions of each of the satellites for day 2.

TABLE 11.1	PROPERTIES OF THE GALILEAN SATELLITES	
Name	Orbital Radius (km)	Orbital Period (days)
Io	4.2×10^5	1.77
Europa	6.7×10^5	3.55
Ganymede	10.7×10^5	7.15
Callisto	18.8×10^5	16.69

Design Your Own Experiment

1. The gravity of Jupiter has pulled the satellites into its equatorial plane, meaning the moons are basically in line with Jupiter's equator. At times the satellites with larger orbital radii appear closer than those with smaller radii. In the previous experiments, you modeled a bird's-eye view of Jupiter's moons, but that's not how we see them from Earth. Find a way to model the edge-on locations of the moons as we do see them from Earth. Since we on Earth are almost perfectly aligned with the orbital planes of the moons, they seem to line up in a straight line across Jupiter's equator. One way to show this is to lay a sheet of clear plastic over the diagram of the positions of the satellites in the previous experiment. Make a red grape-size ball of modeling clay to represent Jupiter. Place this clay ball on the part of the plastic covering Jupiter. Using four different colors of clay, make four pea-size clay balls representing Jupiter's moons and place one ball on dot 1 of each moon's orbit. Slip a piece of stiff cardboard or a book under the paper with the plastic on top, and raise the cardboard so the plane of the plastic is perpendicular to your face and Jupiter is directly in front of your eyes. Observe the positions of each of the satellites in relation to Jupiter. Two diagrams can be made as in Figure 11.2, one a bird's-eye view of the satellites as seen from above Jupiter's north pole (above the plastic) and the other the edge-on view (as seen by an observer on Earth). Repeat this procedure, moving the satellites to their location after 1 Earth day and then again after 2 and 3 Earth days.

Figure 11.2

2. Observe Jupiter through a telescope for a period of 17 or more days. During this time, most of the satellites will have made many revolutions around Jupiter, but Callisto will have made only one. Design an experiment to plot the position of the satellites from day to day. Make four or more observations at least 1 hour apart to identify the satellites. They can be identified by the rate at which they move: from fastest to slowest, they are Io, Europa, Ganymede, and Callisto. For information on identifying the satellites on any specific day, see astronomy magazines, such as *Sky and Telescope* and *Astronomy*, for the month of the observation.

Get the Facts

While the term satellite refers to any object orbiting a celestial body, *planetary satellite* generally refers specifically to moons. The moons of Jupiter are called Jovian satellites. How many Jovian satellites are known to date? How do their distances and orbital periods compare? See an astronomy text and/or search the Web for information about Jovian satellites.

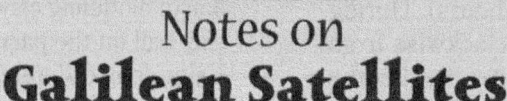

Notes on
Galilean Satellites

Key Facts:

_____

_____

_____

_____

My Results:

_____

_____

_____

_____

_____

_____

_____

_____

Conclusions:

_____

_____

_____

_____

_____

Results of Try New Approaches:

_____

_____

_____

_____

_____

Notes on Designing My Own Experiment:

_____
_____
_____
_____
_____
_____
_____
_____
_____
_____
_____
_____
_____
_____
_____
_____
_____
_____
_____
_____
_____
_____
_____

Notes on Get the Facts:

_____
_____
_____
_____
_____
_____
_____

12 Meteors
Streaks of Light in the Sky

What are those points of light moving across the sky? If they blink or glow red, they are probably on an airplane. A white, continuous, slowly moving light is probably an artificial satellite. The bright streaking white light is a "shooting star," which is actually a meteor.

In this project, you will determine why meteor showers occur by modeling the plane of Earth's orbit in relation to that of a comet. You will discover the two types of meteoroid streams and learn how to represent them. You will also compare the number of meteors in "showers" with the numbers of meteors that fall to Earth sporadically.

Getting Started

Purpose: To model Earth's orbit passing through a comet's orbit.

Materials

pen
1-by-3-inch (2.5-by-7.5 cm) strip of poster board
one-hole paper punch

three 12-inch (30-cm) pipe cleaners— 2 yellow, 1 green

Procedure

1. Draw the Sun in the center of the strip of poster board.

2. Punch four holes in the strip, centering them near the edges of the four ends.

3. Thread the green pipe cleaner through the holes on the short ends of the strip. Twist its ends together, and form it into an elliptical shape. This is a model of Earth's orbit.

4. Twist the ends of the yellow pipe cleaners together to form one long piece. Thread it through the remaining holes. Twist the ends together and form it into an ellipse. This is a model of a comet's orbit.

5. Position the two orbits at angles so that Earth's orbit touches the comet's orbit at only one point (see Figure 12.1).

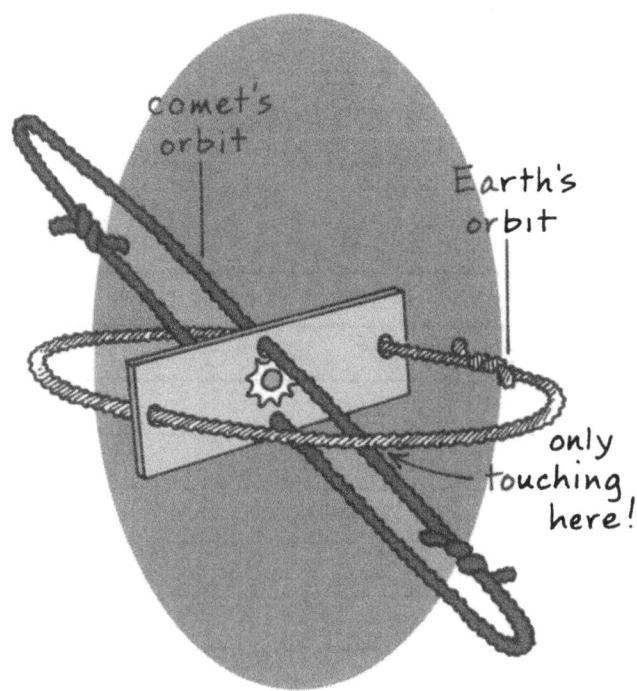

Figure 12.1

Results

The comet's orbit is at an angle nearly perpendicular to Earth's orbit, so they only touch at one point.

Why?

Comets are celestial bodies made of dust, rock, sand, gases, and ices (mainly water and carbon dioxide) that move in an extremely elongated orbit about the Sun. They are often called dirty snowballs. When they get near the inner planets, the Sun's heat vaporizes surface ice, freeing some of the rock, sand, and dust. These pieces become **meteoroids** (solid particles that have broken off a celestial body and that orbit the Sun) scattered along the comet's path. Meteoroids from comets are generally the size of dust specks, while those from **asteroids** (irregularly shaped rocky chunks of matter that rotate as they orbit the Sun, generally between Mars and Jupiter; also called **minor planets**) can be very large.

When a meteoroid enters Earth's atmosphere, it is called a **meteor.** Meteors move so fast that friction with the air causes them to heat up and **vaporize** (change to a gas). The hot vapor is **incandescent** (glowing because of heat). This incandescence is the light of a "shooting star." Any meteor large enough to survive its fall through Earth's atmosphere and actually make it to Earth's surface is called a **meteorite.**

The orbit of Earth and the orbits of most comets never meet. But some comets have orbital paths that do intersect with Earth's orbit. Thus, in its annual trip around the Sun, Earth passes through the orbital paths of several comets at different times throughout the year. Each time this happens, Earth passes through debris the comet left in its path, and a **meteor shower** (an increase in the rate of meteors) is seen by observers on Earth. On any clear night 3 to 15 meteors are generally seen per hour. But during a meteor shower as many as 60 meteors per hour have been seen.

Try New Approaches

Since comets approach the Sun from all directions, most of them are on a different orbital plane than Earth. Therefore, Earth generally passes through the path of any given comet only once each year. However, it is possible for Earth to intersect a comet's path twice each year—once where the comet is coming in toward the Sun, and again when it is heading back out away from the Sun. This is the case with the famous Halley's comet, which happens to have an orbital plane similar to Earth's. It produces two annual meteor showers: the Eta Aquarids in May and the Orionids in October. Model this by repeating the experiment, making the four holes in your poster board in line with one another, with two holes on each side of the Sun.

Design Your Own Experiment

1a. The direction of most meteors is sporadic. The remains of a long-dead comet are spread widely across space. Thus meteoroids from these comets may enter Earth's atmosphere from any and all directions. Design an experiment that compares the number of meteors coming from one direction during a shower to the numbers that come sporadically during a given time period. One way is to make nightly counts during a meteor shower. Do the same when the meteor fall is sporadic. For information about how to observe and count meteors, as well as dates of expected meteor showers, see Fred Schaaf's *40 Nights to Knowing the Sky* (New York: Holt, 1998), pp. 120–127.

b. The meteoroids of a meteor shower are part of a **meteoroid stream** (debris from a comet traveling in or near a comet's orbital path). These meteoroids may be bunched, or concentrated, in one part of the comet's orbit. When Earth passes through such a bunched swarm of meteoroids, observers on Earth are likely to see a **meteor storm** (many meteors occurring in a short period of time). Or the meteoroids may be rather uniformly distributed along a comet's path. This occurs when comets have made many trips around the Sun, and generally produces more sparse meteor showers. Design a display for these two types of meteoroid streams.

Get the Facts

Meteors sometimes leave a glowing trail across the sky called a *meteor train.* What causes this trail? How long does it last? For information about meteor trains and other types meteors, such as fireballs and bolides, see Schaaf's *40 Nights,* pp. 118–120.

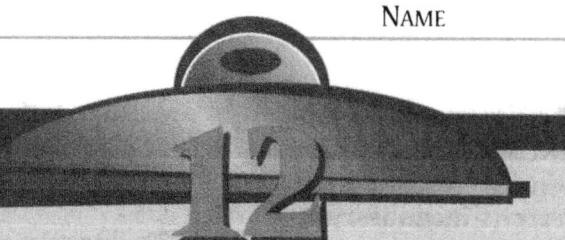

Notes on
Meteors

Key Facts:

_____
_____
_____
_____

My Results:

_____
_____
_____
_____
_____
_____
_____

Conclusions:

_____
_____
_____
_____
_____

Results of Try New Approaches:

_____
_____
_____
_____
_____

Notes on Designing My Own Experiment:

_____
_____
_____
_____
_____
_____
_____
_____
_____
_____
_____
_____
_____
_____
_____
_____
_____
_____
_____
_____
_____
_____
_____
_____
_____

Notes on Get the Facts:

_____
_____
_____
_____
_____
_____

Biology

13 Meristematic Region
Growth Zone

Dicots, the popular name for dicotyledons, are one of the two large groups of flowering plants. In dicots the seeds have two cotyledons, which is the leaflike part of the seed containing nutrition for the seed to develop and grow. All types of plants contain areas of rapidly growing tissue, called meristem. The meristem is responsible for growth, both in the length and the width of plant stems.

In this project, you will discover the location of the meristematic region in dicot plant stems responsible for the elongation of the stems. You will discover if there is a change in meristematic growth rate during different times of the day and among different plants. You will determine any effect of light duration on meristematic growth. You will also research how the primary and secondary growth of a plant affects its size.

Getting Started

Purpose: To determine where growth occurs in the stem of a pinto bean plant.

Materials

three 9-ounce (270-ml)
  paper cups
potting soil
ruler

pencil
12 dry pinto beans
tap water
marking pen

Procedure

1. Fill each cup with potting soil to within 2 inches (5 cm) from the top.

2. Use the pencil to punch six or more holes near the bottom of each cup.

3. Lay four beans on the surface of the soil of each cup.

4. Cover the beans with about 1 inch (2.5 cm) of soil.

5. Moisten the soil with water. Keep the soil moist but not wet throughout the experiment.

6. Allow the seeds to germinate and the seedlings to grow to a height of about 6 inches (15 cm)

above the top of each cup. *Note:* This will take seven or more days.

7. Select 4 or more of the most healthy looking seedlings.

8. Use the marking pen to draw a line across the stem of each selected plant level with the lip of the cup. Use this mark as the starting line.

9. Mark three equal sections on each stem between the cotyledons (the two remaining parts of the seed) and the starting line (see Figure 13.1).

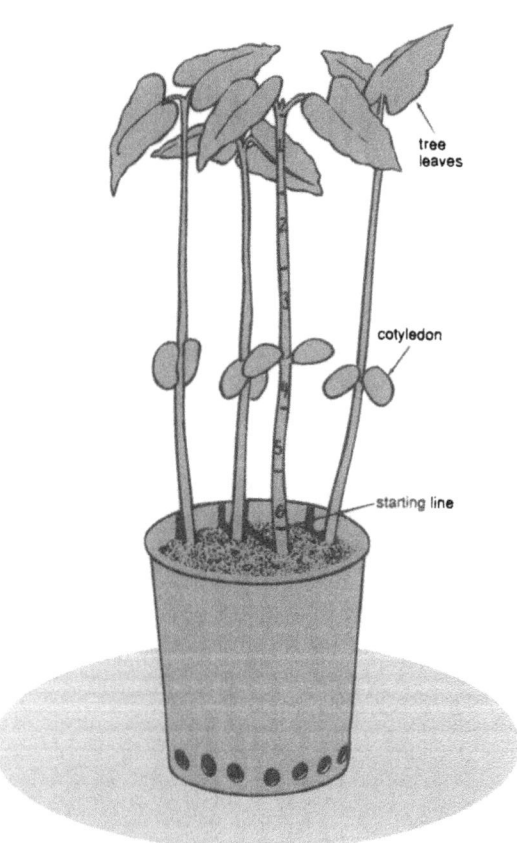

Figure 13.1

10. Mark three equal sections on each stem between the cotyledons and the true leaves at the top of the stem.

11. Prepare drawings of each plant, numbering each section from 1 to 6, starting at the top of each stem as in Figure 13.1.

12. Measure and record the starting length of each section in a Stem Growth Data table, such as the one in Table 13.1.

13. Set the plants near a window with indirect sunlight. Add equal amounts of water to each cup to keep the soil moist but not wet. After seven days, again measure and record the length of each section. Determine any change in the length of each section.

TABLE 13.1 STEM GROWTH DATA—PLANT 1

	Section Length, mm					
	1	2	3	4	5	6
Start						
Final						
Difference						

Results

The author's plants showed the greatest elongation in the sections marked 1, only slight changes in the sections marked 2, and no measurable changes in the lower sections marked 3 through 6.

Why?

New plant cells are formed at growing points of actively dividing cells. These growing points are called **meristems.** The specialized clusters of cells in the meristem produce growth by mitosis. In **mitosis,** the cell parts duplicate themselves and then divide into two separate cells. With each division at least one of the new cells, called daughter cells, remains meristematic, thus preserving the meristem. The second daughter cell may be meristematic for a time, but eventually forms a specialized cell.

A bean plant is a **dicotyledon (dicot),** which is one of two groups of flowering plants, characterized by having seeds with two cotyledons. In a dicot, the **cotyledons** are the two parts of the seed, also called seed leaves, that contain stored **nutrients** (nourishing materials needed for life and growth of **organisms**—living things) required for the seed to grow and develop. The **stems** (part of a plant that supports other plant parts, including leaves, flowers, and fruit) of the bean plants in this experiment have meristems located at the **apex** (tip) of their stem, which are called **apical meristems.** Division of apical meris-

tem cells increases the length of the stem, at the top of the stem.

Try New Approaches

1. Does the rate of meristematic growth in the bean plants change during different periods of the day? Repeat the experiment marking only the three sections of the stems above the cotyledons. Measure these sections at two-hour intervals during the day, starting early in the morning and stopping at sunset.

2. Does the apical meristem growth rate differ in different dicot plants? Repeat the original experiment using different kinds of bean seedlings such as pinto, lima, and red kidney beans. For a better comparison of the growth, divide the area above the cotyledon of each plant into four or more sections. **Science Fair Hint:** Use the bean seedlings as the control so that growth of the other seedlings is compared to the bean seedlings. Use diagrams showing the changes in the length of each section as part of a project display.

Design Your Own Experiment

1. The interval in a 24-hour period in which an organism is exposed to light is called a **photoperiod** (length of daylight and darkness). The responses of organisms to changes in the photoperiod is called **photoperiodism.** Design an experiment to determine if light duration affects the rate of meristematic growth. One way is by repeating the original experiment using three test groups of plants with 3 cups of plants in each group. Make every effort to vary only the amount of light that each group receives. Loosely cover the cups in one group with clear cellophane, taping the edges and top of the cellophane together to form a transparent (allows light to pass through) tubelike "hat" over each cup. Use wax paper to prepare a **translucent** (allows some light to pass through, though this light is **diffused**—spread out in different directions) "hat" for the second group, and aluminum foil to prepare an opaque (doesn't allow light to pass through) "hat" for the third group. Let the cups with the opaque hats be the control group. Lift the hats and measure the plants daily.

2. Plants grow in different environments with different nutrients. Design an experiment to determine how nutrients affect meristematic growth. One way is to repeat the original experiment using three sets of plants with the independent variable being nutrient. You will need to determine a method of growing the plants in soils from three different regions. Take care in the design of the experiment to keep all factors constant except for the independent variable, soil, for each set of plants the same.

Get the Facts

1. Some plants live for hundreds and even thousands of years. Some bristlecone pines found in eastern California have been documented as being many thousands of years old. Theoretically these plants can grow forever by substituting new cells for aging and nonfunctioning cells. This potential for continuous growth is called *open growth* and involves apical and secondary meristems. Growth from apical meristems is called *primary growth* while growth from *lateral meristems* found in the *cambrium* (growth layer in plant stems) is called *secondary growth*. How does primary and secondary growth affect the size of a plant? What produces growth rings in some plants? Do all plants have cambrium? For more information, see a biology text.

2. Some plants have a main stem that grows taller than other stems on the plant. This is called a *terminal shoot*. What causes a terminal shoot? What is apical dominance? For more information, see *Janice VanCleave's A+ Projects in Biology* (New York: Wiley, 1993), pp. 57–62.

Notes on
Meristematic Region

Key Facts:

_____

_____

_____

_____

My Results:

_____

_____

_____

_____

_____

_____

_____

_____

Conclusions:

_____

_____

_____

_____

_____

Results of Try New Approaches:

_____

_____

_____

_____

_____

Notes on Designing My Own Experiment:

Notes on Get the Facts:

14 Seed Parts

Exterior and Interior

Given the right amount of water, oxygen, and warmth, most seeds germinate and develop into mature plants. Seeds vary in physical appearance both on the outside and on the inside.

In this project you will have the opportunity to identify exterior and interior seed parts. The seed parts of two basic seed types, dicotyledons and monocotyledons, will be compared. You will determine why the cotyledon is necessary and investigate the presence of starch in each seed part.

Getting Started

Purpose: To dissect a bean seed and identify its parts.

Materials

8 to 10 pinto beans
1-pint (500-ml) jar
distilled water
refrigerator

paper towel
magnifying lens (handheld type)

Procedure

1. Place the beans into the jar and cover them with distilled water.

2. Put the jar of beans in the refrigerator to reduce bacterial contamination.

3. Soak the beans for 24 hours.

4. Remove the beans from the jar and place them on the paper towel to absorb the excess water.

5. Inspect the outside of the beans and identify the seed coat and hilum (see Figure 14.1).

6. Use your fingernails to carefully remove the seed coat from one of the beans.

7. Very gently pry the rounded side of the bean open like a book with your fingernail.

8. Spread open the two halves of the bean.

9. Use the magnifying lens to study the inside of each bean half and identify the cotyledon,

plumule, epicotyl, hypocotyl, and radicle (see Figure 14.1).

10. Open several beans and compare their parts for differences in size, shape, and organization.

Figure 14.1

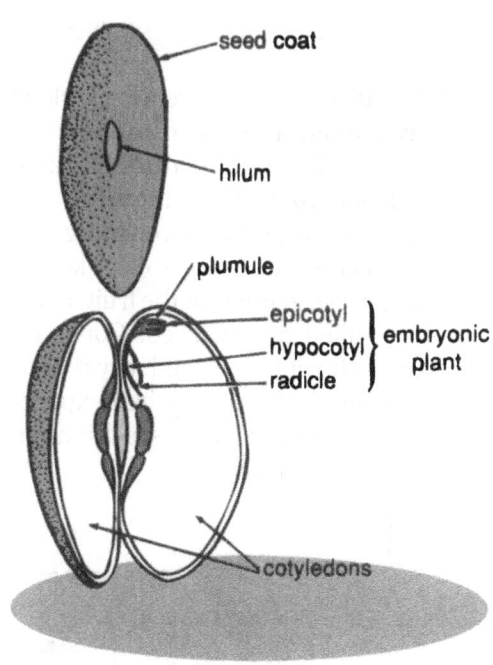

Results

The brown seed coat is thin and peels off easily to reveal a white structure with two separate halves connected at the top. Inside is a cylindrical structure with what appears to be folded leaves at the end.

Why?

The bean seed consists of three basic parts: a seed coat, an embryo, and two cotyledons. The surface covering of a seed is called the **seed coat.** This jacket around the seed protects the **embryo** (an organism in its earliest stage of development) from insects, disease, and other damage. The **hilum** is the scar on the seed coat of some seeds, such as beans; it is the point where the seed was attached to the plant during development.

A bean has two halves under the seed coat called cotyledons, so it is called a dicotyledon or a dicot (a plant whose seed has two cotyledons). The end of the embryo that develops into the **shoot** (part of plant that grows above ground) is called the **epicotyl.** The tip of the epicotyl that looks like folded leaves is called the **plumule** (embryonic shoot tip that consists of several tiny, immature leaves that at maturity form the first true leaves). Below the epicotyl is the **hypocotyl,** which attaches the embryo to the cotyledon. The end of the hypocotyl, called the **radicle,** develops into the root.

Try New Approaches

1a. While dicotyledons have two cotyledons, **monocotyledons,** also called **monocots,** have only *one* cotyledon. How do the seed parts of a monocotyledon compare to those of a dicotyledon? Repeat the experiment replacing the bean seed with corn seed. *Note:* A corn kernel is a fruit. The corn seed is inside the fruit, but the kernel is often called a corn seed. Corn seed is much more difficult to open and may require the use of a scalpel. As in the original experiment, study and identify seed parts (see the corn kernel diagram in Figure 14.2).

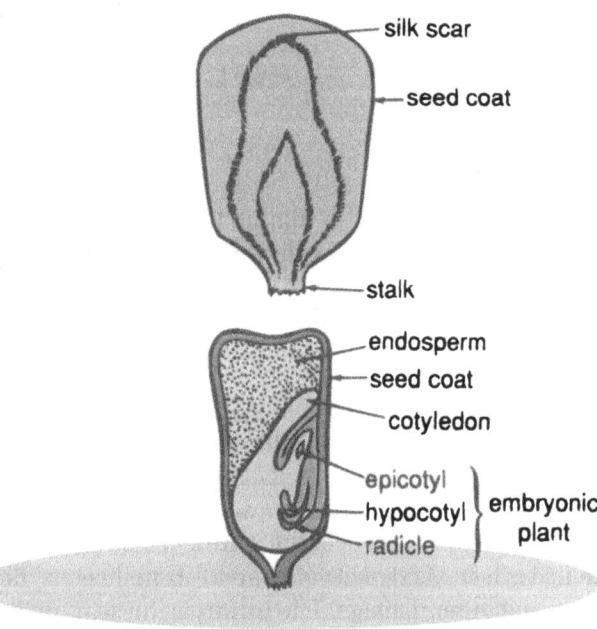

Figure 14.2 Corn Kernel

CAUTION: *Adult supervision is needed when cutting with a sharp instrument. Always cut in a direction away from your hands and fingers.*

b. Seeds vary in size, shape, and color. **Science Fair Hint:** Collect and display samples of seeds. Organize the seeds to indicate which are **angiosperms** (flowering seed plants) and which are **gymnosperms** (a nonflowering seed plant). Further classifications, such as how the seeds are transported, can also be used. For information about seeds see Cynthia Overbeck's *How Seeds Travel* (Minneapolis: Lerner Publications Company, 1990).

2. Use a **stereomicroscope** (dissecting microscope that allows thick specimens to be observed) to study, identify, and compare the parts of bean seed and corn seed. **Science Fair Hint:** Diagram the parts of each seed type, label the parts, and display the diagrams.

Design Your Own Experiment

1. Bean seeds have two cotyledons. Are both cotyledon's necessary for their **germination** (process by which a seed grows)? One way to determine this is to grow bean seeds with different amounts of the two cotyledons. Soak 30 beans for 24 hours in a jar of water. Prepare five or more beans for each of the following five groups:

- 100%—Do not cut away any part of the cotyledons.

- 75%—Cut away the lower half of one cotyledon.

- 50%—Cut away the lower half of both cotyledons.

- 25%—Cut away all but one-fourth of the cotyledons, leaving the section attached to the embryo.

- 0%—Cut away both cotyledons, leaving only the embryo.

Note: Use a scalpel to carefully cut away the indicated parts of the cotyledons (see Figure 14.3). Make every effort not to disturb the embryo. Discard extra beans.

Place the prepared beans on a moist paper towel. Set the towel on a sheet of aluminum foil and fold the foil around the beans. Open the foil package daily and make observations of any evidence of germination. **Science Fair Hint:** Use diagrams and photographs to represent results.

2. Starch is a stored food supply for seeds. Is starch stored in seed parts other than the cotyledons? One way to determine this is to test for the presence of starch using iodine. **CAUTION:** *Keep iodine out of the reach of small children. It is poisonous and is for external use only. It can stain skin, fabric, and furniture.* A positive test for the presence of starch is a dark blue-black color on the material where the iodine is placed. Using bean seeds, determine the presence of starch by using iodine on each of the following seed parts: cotyledon, epicotyl, hypocotyl, radicle, and seed coat. Separate the parts from one another and place them in shallow containers. Add one or two drops of tincture of iodine (found at a pharmacy) to each part. Prepare and display a data table to indicate the presence of starch. Use a plus sign (+) to indicate the presence of starch and a minus sign (–) to indicate its absence.

Get the Facts

1. *Endosperm* (the food-storage tissue of a seed) is present at some stage during the development of all seeds. It is present in mature corn seed but not in bean seed. What happens to the endosperm? Why is it present in some mature seeds but not in others?

2. The seed coat of different seeds varies in color, thickness, and texture. Sometimes, it is smooth and paper thin, as on a bean. A coconut's seed coat is rough, thick, and hard. This outer coat protects the embryo from drying out against injury from falls or being struck by objects, and from attacks by insects, some animals, bacteria, and fungi. It also insulates the embryo from extreme temperatures. Germination cannot take place unless the seed coat is cracked. The seed coats of some plants are cracked by alternate freezing and thawing. Not all seeds are affected by temperature changes, and many plants do not live where the temperature drastically rises and falls. How else might a seed coat crack, allowing water and oxygen in and the developing embryo to emerge?

3. Many types of seeds do not germinate regardless of the environmental conditions. They germinate only after a period of rest called the *dormancy period*. Dormancy may be the result of many factors; for example, the seed coat may contain chemical inhibitors that prevent germination. During a period of rest soil moisture leaches out the chemicals or they break down as they react with other chemicals in the soil. When the inhibitors are gone, the seed germinates. Find out more about seed dormancy and the events that result during a rest period.

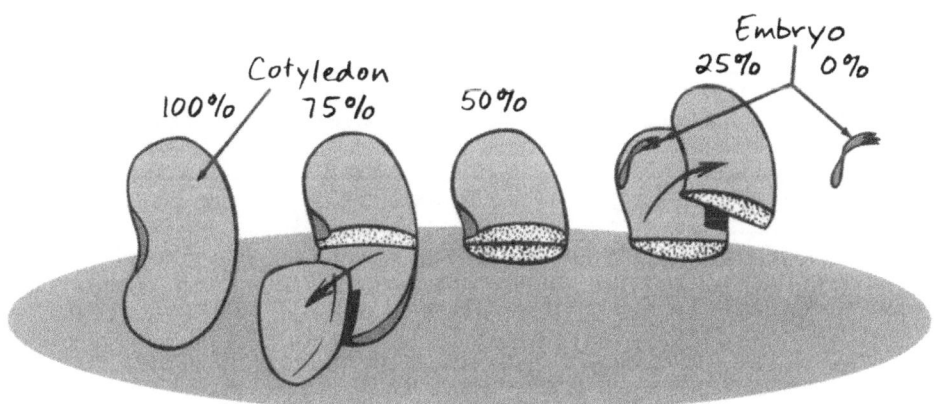

Figure 14.3

NAME

Notes on
Seed Parts

Key Facts:

_____

_____

_____

_____

My Results:

_____

_____

_____

_____

_____

_____

_____

_____

Conclusions:

_____

_____

_____

_____

_____

Results of Try New Approaches:

_____

_____

_____

_____

_____

Notes on Designing My Own Experiment:

_____

_____

_____

_____

_____

_____

_____

_____

_____

_____

_____

_____

_____

_____

_____

_____

_____

_____

_____

_____

_____

_____

_____

_____

Notes on Get the Facts:

_____

_____

_____

_____

_____

Autotrophs
Food Builders

Respiration is the chemical and physical process by which animals and plants take in oxygen and use it to release energy from glucose molecules. Respiration provides energy necessary for carrying on all essential life processes. Plants are autotrophic—that is, they are able to manufacture glucose through a process called photosynthesis. Photosynthesis is a process in which energy from light is captured and stored in the plant.

In this project, you will use plants to examine the processes of photosynthesis and respiration and determine substances produced and consumed in each. You will study the effect of light on these processes. You will also determine if respiration occurs during the germination of seeds.

Getting Started

Purpose: To test for the production of carbon dioxide by respiration in plants.

Materials

five 6-inch (15-cm) test tubes
distilled water
masking tape
pen
ruler
four 5-inch (15-cm) sprigs of elodea or other water plant the length of the test tube (found at a pet store)

0.04% aqueous solution of bromthymol blue indicator (your teacher can purchase this from a scientific supply company)
5 corks to fit the test tubes
aluminum foil
bowl or container that will hold the test tubes upright

Procedure

1. Rinse the test tubes with distilled water.
2. Place a piece of tape about 2 inches (5 cm) long from the top edge down the side of each tube. Use the pen and ruler to mark a line across the tape 1 inch (2.5 cm) from the top of each tube.
3. Use the pen to label four of the test tubes "1," "2," "3," "4." The fifth tube will be the control. Label this remaining tube "C."

4. Place an equal-size sprig of elodea into test tubes 1 through 4.
5. Fill all the test tubes including C up to the 1-inch (2.5-cm) line with bromthymol blue indicator. Put a cork in each of the tubes.
6. Observe and record the color of the liquid in each tube in a Carbon Dioxide Data table like the one shown in Table 15.1.

Figure 15.1

	Color of Liquid in Test Tubes				
Time, hr	Test tube 1	Test tube 2	Test tube 3	Test tube 4	Test tube C
0					
↓					
24					

TABLE 15.1 CARBON DIOXIDE DATA

7. Cover the test tubes with aluminum foil so that no light can enter.
8. Stand the tubes together in a bowl and place the bowl on a table in an unlighted or dimly lit area.
9. Observe the color of the liquid in each test tube every hour for eight or more hours during the day, then again after 24 hours. Record the color in the data table.

Figure 15.2

Results

The color of the liquid in test tube C remains unchanged and is blue. The color of the liquid in test tubes 1 through 4 changes from blue to green to yellow as time passes.

Why?

Indicators are natural or synthetic chemical substances that change color in response to other chemicals. **Acid indicators,** such as bromthymol blue, change color in the presence of a substance called acid. **Acids** cause the bromthymol blue indicator to turn from blue to green, then to yellow, depending on the **concentration** (amount of one substance dissolved in a given volume of another) of acid present. At a low acid concentration the color is green and at a higher concentration the color is yellow.

A **chemical reaction** is a change that produces one or more new substances. In a chemical reaction, a **reactant** is a substance that is changed and a **product** is a substance that is produced. **Internal respiration** involves a series of chemical reactions that liberate energy from **glucose** (a type of sugar) molecules. Respiration is a catabolic process that occurs inside every animal and plant cell. A **catabolic process,** is a chemical reaction in which larger substances in an organism are broken down into smaller ones. During respiration the catabolic process is the breakdown of a glucose to water and a gas, carbon dioxide, with a release of energy. Internal respiration is also called **aerobic respiration,** which means that oxygen is one of the reactants. The general equation for aerobic respiration is:

glucose + oxygen→water + carbon dioxide + energy

The products of respiration of the elodea plant (carbon dioxide and water) mix with the liquid indicator. Some of the carbon dioxide chemically combines with the water, forming the acid called carbonic acid. The more carbon dioxide that dissolves in the water over time, the greater the concentration of the acid formed. The color of the liquid in each test tube indicates changes in the concentration of acid as a result of the respiration of each elodea plant over time.

Try New Approaches

1. Respiration occurs with or without the presence of light, but light is necessary for photosynthesis. **Photosynthesis** is the process by which light energy is absorbed and then converted to the chemical energy of glucose. This conversion occurs in plants that have chlorophyll (pigment that absorbs light). Photosynthesis is an **anabolic process,** in which energy is needed to build larger molecules from smaller ones. During photosynthesis, carbon dioxide and water in the presence of chlorophyll and light energy combine, forming glucose and oxygen. The general equation for photosynthesis is:

$$\text{carbon dioxide + water} \xrightarrow[\text{light energy}]{\text{chlorophyll}} \text{glucose + oxygen}$$

Determine how light affects the concentration of carbon dioxide released by plants into the environment around them. Do this by repeating the experiment but leave off the aluminum foil covering the test tubes. Then place the test tubes 8 inches (20 cm) from a desk lamp. Compare the color of the test tubes in the light with those from the previous experiment that did not receive light.

2. Does the amount of light affect the results? Repeat the previous experiment twice by placing one set of uncovered test tubes closer to the light and a second set of test tubes farther away from the light.

3. Does the source of light affect the results? Repeat the experiment by placing five uncovered test tubes near different light sources such as sunlight, incandescent bulbs, fluorescent bulbs, and plant grow-lights. When using artificial lights, place the test tubes an equal distance from the lights.

1. Seeds do not need light to germinate. Design an experiment to find out whether respiration occurs during seed germination. One method is to fill a small plastic soda bottle one-third full with dry pinto beans and cover the beans with distilled water. Place the jar in a refrigerator to reduce the production of bacteria while the beans soak. (Soaking the beans allows water to penetrate them, which encourages germination.) After soaking overnight, pour off the water and use modeling clay to secure one end of a flexible plastic drinking straw inside the mouth of the bottle. Insert the free end of the straw into a glass of bromthymol indicator (see Figure 15.3). Observe the color of the indicator solution near the end of the straw to determine whether carbon dioxide is exiting the straw.

bromthymol blue

Figure 15.3

2a. The chemical equations for photosynthesis and aerobic respiration are:

Photosynthesis:

$$6\ CO_2 + 6\ H_2O \xrightarrow[\text{Light Energy}]{\text{Chlorophyll}} C_6 H_{12} O_6 + 6\ O_2$$

Aerobic Respiration:

$$C_6 H_{12} O_6 + 6\ O_2 \rightarrow 6\ CO_2 + 6\ H_2O + \text{Energy}$$

While photosynthesis requires light, aerobic respiration does not. Note that the products of one of the energy equations are the reactants for the other. Plants are **autotrophs,** which are organisms that can make their own food. So plants can live in a closed environment as long as they have access to sunlight, because they can recycle matter (transforming sugar and oxygen into carbon dioxide and water, and back again). They do need access to sunlight, however, because light energy cannot be recycled. Design an experiment to determine how long a plant can live in a closed container without any outside nutrient sources. One way is to prepare eight jars of distilled water and equal amounts of elodea sprigs. Use lids to close four of the jars. Then set all of the jars near a window with indirect sunlight or in a well-lighted area of a room. Observe the plants for one month or more. Compare the plants in the covered jars with those in the uncovered jars.

b. The previous experiment can be performed by filling the jars about one-fourth full with soil. Use plants such as grass or ivy. Moisten the soil in each jar, then do not add any nutrients.

Get the Facts

1. *Cellular respiration* is the process by which *adenosine triphosphate, ATP* (a special energy-carrying molecule) is produced by respiration. How is ATP involved in energy production? What part do ADP molecules play in the production of ATP molecules? How much ATP is produced during aerobic respiration? What happens to the ATP molecules? Where does aerobic respiration take place in living organisms? For information about respiration, see a biology textbook.

2. Photosynthesis is said to require light, but it actually involves two processes, one called the light reaction and the other the dark reaction. What is the difference between these two reactions? What is needed for these reactions to occur? A biology text can provide information about photosynthesis.

Notes on
Autotrophs

Key Facts:

_____

_____

_____

_____

My Results:

_____

_____

_____

_____

_____

_____

_____

Conclusions:

_____

_____

_____

_____

_____

Results of Try New Approaches:

_____

_____

_____

_____

_____

Notes on Designing My Own Experiment:

Notes on Get the Facts:

16 Hydroponics
Growth without Soil

Plants need nutrients for survival. Terrestrial plants, which grow in the ground or soil, have roots to gather dissolved nutrients from the soil. But do the nutrients have to come from the soil?

In this project, you will observe hydroponic growth—that is, the growth of plants in a nutrient solution without soil. The effects of sunlight, the amount of oxygen, and the growing medium will be studied. You will also compare plants grown in soil to those grown in liquid nutrients.

Getting Started

Purpose: To construct a **hydroponicum** (hydroponic growing unit) for tomato plants.

Materials

5-gallon (20-liter) bucket	masking tape
water	3 dwarf tomato bedding plants
1-gallon (4-liter) plastic milk jug with cap	scissors
1 tablespoon (15 ml) of plant fertilizer (5–10–5)	3 7-oz (210-ml) paper cups
1 teaspoon (5 ml) of Epsom salt	roll of paper towels
1 teaspoon (5 ml) of household ammonia	3 1-pint (500-ml) glass jars
marking pen	3 1-x-1-foot (30-x-30-cm) sheets of aluminum foil

Procedure

1. Fill the bucket with tap water. Allow the water to stand in this open container for one day so that the chlorine in the water can evaporate.

2. Prepare the nutrient solution as follows:

   a. Fill the milk jug one-fourth full with the dechlorinated water from the bucket

   b. Add the plant fertilizer, Epsom salt, and household ammonia to the water in the jug. **CAUTION:** Ammonia is a poison. It and its fumes can damage skin and mucous membranes of nose, mouth, and eyes.

   c. Secure the cap and vigorously shake the jug to dissolve the solids.

   d. Add more water from the bucket to fill the jug to within 2 inches (5 cm) from the top. Rotate the jug back and forth to mix.

3. With the marking pen, write "Nutrient Solution" on a piece of masking tape and tape this label to the jug.

4. Remove one tomato plant from its bedding pot.

5. Set the plant in the bucket of dechlorinated water and gently move it back and forth to remove as much of the soil from the roots as possible.

6. Use scissors to cut a hole in the bottom of one paper cup just large enough so that the tomato plant's roots fit through. You want the root to fit snugly in the hole with about one-fourth of the root's length remaining inside the cup.

7. Stuff pieces of paper towels inside the cup to give the plant support.

8. Set the paper cup inside the neck of the jar to determine how far the bottom of the cup sits in the jar.

9. Remove the paper cup and add enough nutrient solution so that its surface will be just beneath the bottom of the paper cup. Keep the level of the liquid the same throughout the experiment.

10. Reposition the paper cup inside the jar (see Figure 16.1 on the following page)

11. Surround the jar with one sheet of aluminum foil to prevent algae from growing in the liquid.

12. Repeat this procedure (steps 4 through 11) for the remaining plants.

13. Place the hydroponicums near a window in direct sunlight.

14. Keep a daily record of the growth of the plants for as long as it takes for the plants to mature and bear fruit.

15. Place the plants in larger containers as their growth requires.

tomato plant

paper towel

paper cup

glass jar surrounded
with aluminum foil

roots

nutrient solution

Figure 16.1

Results

Results will vary, but healthy potted plants can mature and bear fruit in a hydroponic environment.

Why?

Growing plants in a 100% liquid nutrient solution without a supporting medium is called **aquaponics.** In this experiment, the paper products provide support for each plant's root and stem system, but they do not add nutrients to the plant. As long as the medium in which the plant is grown does not add nutrients and all the nutrients derived from a nutrient solution, the method of growing is considered to be hydroponic. The hydroponic growing unit, a hydroponicum, fits this description.

Try New Approaches

1. Does the amount of sunlight received by a tomato plant affect its growth? Repeat the procedure to prepare six hydroponicums of tomato bedding plants. Place two plants in each of the following situations: all day sun, morning sun, and evening sun. You may have to use shields made of boards to block the sun from directly hitting the plants that are to receive only partial sunlight (see Figure 16.2). Keep daily records. Measure the height of each plant, count the leaves, and note the general appearance of the plant's leaves and stem. **Science Fair Hint:** Prepare and display a graph to represent the growth of each plant (see the sample graph in Figure 16.3).

2. Does the amount of oxygen received by the roots affect the growth of the plant? Prepare three hydroponicums using tomato bedding plants. Place a small aquarium aerator in each jar. Allow the liquid in the three units to be aerated for different time periods, such as 2, 8, and 24 hours.

3. Can other plants be grown hydroponically? Repeat the original experiment using bedding plants such as strawberries or flowering plants.

Figure 16.2

4. Does a growing medium affect the results? A plant can be considered hydroponically grown as long as the medium does not provide nutrients to the growing plant. Plant a tomato bedding plant in a nonnutrient medium such as LECA (light expanded clay aggregate), sand, gravel, charcoal, sawdust, and vermiculite or perlite. You could choose to use several mediums and compare their results. Use the prepared liquid nutrient from the original experiment.

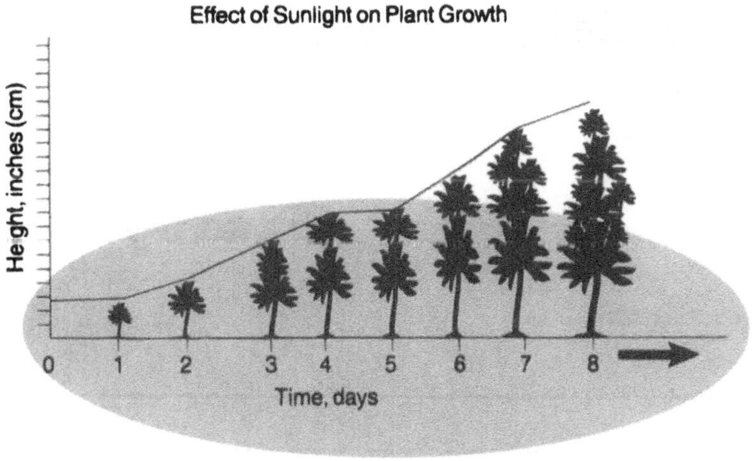

Effect of Sunlight on Plant Growth

Figure 16.3

Design Your Own Experiment

If the same nutrient additive is used, do terrestrial plants grow better than hydroponic plants? Plant three tomato bedding plants in soil. Follow the instructions from a professional at a nursery for the planting and care of the plants. Use the same nutrient solution to water both the plants in the hydroponicums and those in the soil. Take photographs of the different stages of growth of each plant, noting the date and time you take each photograph. Be sure to include pictures of the fruit grown from the plants. Use the sample plant growth graph (Figure 16.3) to prepare graphs for each plant. Display the photographs to represent procedure, steps and results. Display several of the graphs to show the comparison between geoponic (grown in the earth) and hydroponic growth.

Get the Facts

1. What nutrients are needed for proper plant growth? Use a gardening book to find the elements required for plant growth and their source and functions. What are the symptoms of specific nutrient deficiencies?

2. Seaweed is an example of growth by natural aquaponics. What other plants grow aquaponically in nature? You could display pictures of these plants to show productive aquaponic growth.

3. What are the advantages of hydroponics over geoponics? Find out more about growing plants in liquid instead of soil. Is it more or less expensive? Is it more productive? Is it more environmentally sound?

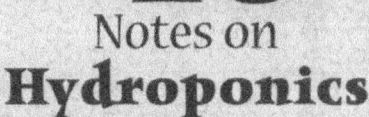

Notes on
Hydroponics

Key Facts:

_____

_____

_____

_____

My Results:

_____

_____

_____

_____

_____

_____

_____

Conclusions:

_____

_____

_____

_____

_____

Results of Try New Approaches:

_____

_____

_____

_____

_____

Notes on Designing My Own Experiment:

_____
_____
_____
_____
_____
_____
_____
_____
_____
_____
_____
_____
_____
_____
_____
_____
_____
_____
_____
_____
_____
_____
_____
_____
_____
_____
_____
_____

Notes on Get the Facts:

_____
_____
_____
_____
_____
_____
_____

17 Apical Dominance
Growth Inhibitor

Certain plant hormones known as auxins are responsible for many plant responses. Auxins affect, for example, the growth of buds, stems, leaves, and roots.

In this project, you will have the opportunity to demonstrate apical dominance, which is the inhibition of lateral bud development by the growth of a terminal shoot—an effect caused by auxins. You will also study the effects of auxins on plant activities such as the stimulation of fruit development and the shedding of leaves and fruit (abscission).

Getting Started

Purpose: To determine whether the terminal bud on a white potato exhibits apical dominance.

Materials

marking pen   baking pan
10 white potatoes

Procedure

1. Use the marking pen to number each potato.

2. Observe and record the appearance of each potato.

3. Place the potatoes side by side on the baking pan. Try to give them as much space as possible.

4. Set the pan in a closed cabinet.

5. Observe and record the appearance of the potatoes weekly until a 6-inch (15-cm) growth of one of the buds is observed. This could take three or more weeks.

6. When observing the bud growth, be very careful not to break any buds. *Note:* Save the potatoes for later experiments.

Results

A bud on the end of the potato grows into a long shoot, but the other bud growth is shorter.

Why?

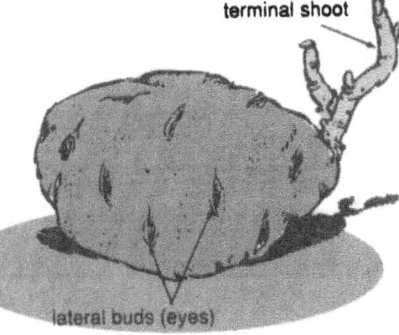

terminal shoot

lateral buds (eyes)

Figure 17.1

A potato is actually part of the underground stem of a potato plant. The buds on a potato are often called "eyes." The eye at the end of the potato that develops into a long shoot is a **terminal** (apical) bud; the remaining eyes are referred to as **lateral** buds because they grow from the side of the stem (see Figure 17.1).

The growth of the terminal bud on the potato and on other plant stems generally inhibits the development of the lateral buds on the stem below. This inhibition of lateral shoot growth because of the presence of a terminal shoot is called **apical dominance.** Apical dominance is more pronounced in plants that have tall, single stems (such as pine trees), but even short, bushy shrubs can develop a single terminal shoot. Gardeners prune the terminal shoots from shrubs and trees to make the plants' lower stems grow. They know that fuller trees and bushes result if the terminal shoots are cut.

Removal of the terminal shoot affects the distribution of auxin (a hormone that causes the cells in a plant to lengthen) in the stem. Apical dominance is thought to be the result of the production of auxin in the meristem cells of the terminal bud that are then transported downward. **Meristem** cells are located at the top of each stem, and it is these cells that divide by a process called **mitosis.** The cell division in the meristem of apical buds is stimulated by the presence of auxin, but the downward transport of auxin produced in the apical buds inhibits cell division in the lateral buds.

Try New Approaches

1. Can lateral buds develop without the presence of a terminal shoot? Repeat the experiment two time, first using slices from potatoes with devel-

oping shoots, and then using slices from potatoes with underdeveloped buds. For the first experiment, use several potatoes from the original experiment that have well-developed terminal shoots and partially developing lateral shoots. Cut ten sections from the potatoes, one shoot to a section. Break off any extra shoots so that there is only one shoot per slice. For the second experiment, use several potatoes not from the original experiment that do not have developing shoots. Cut ten sections from the potatoes, one bud to a section (see Figure 17.2).

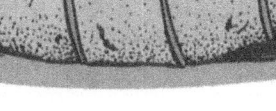

one bud per slice

Figure 17.2

2a. Do lateral buds develop if the terminal bud is removed? Repeat the original experiment. As the buds develop, remove the dominant terminal buds from all but two of the potatoes. These two will be the controls. **Science Fair Hint:** As the buds develop, measure the shoots on the potatoes and use graphs to represent the growth of shoots with and without the terminal shoots.

b. When the terminal shoot is removed, do any of the lateral shoots become more prolific and reimpose apical dominance? As the preceding experiment processes, observe the lengths of the shoots and determine whether one or more shoots on the potatoes begin to outgrow the others and replace the removed terminal shoots (see Figure 17.3).

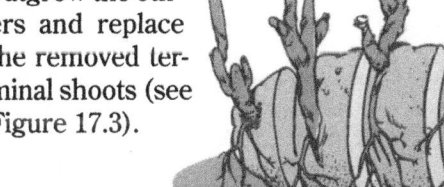

Figure 17.3

Design Your Own Experiment

Does the auxin in commercial products such as blossom set actually prevent fruit from prematurely falling from the plant? Is it possible that the plant would produce its own auxin if provided with the proper nutrients? Grow tomato plants in soil with varying amounts of nitrogen and phosphorous. With the assistance of a professional at a nursery, select at least eight dwarf tomato plants that can be grown in large buckets and choose nitrogen- and phosphorous-rich fertilizers and a prepared product designed to set the blossoms on tomato plants. Divide the plants into two groups. At the time suggested on the blossom-set product label, spray half of the plants in each group. Table 17.1 indicates the different nitrogen (N) and phosphorous (P) levels that each plant receives. It is also a sample data table that you can use when recording and displaying your results.

Get the Facts

1. A chemical growth stimulator called auxin seems to affect fruit development, apical dominance, abscission, positive and negative tropism, and other plant growth. Use biology texts to find out more about the discovery and role of auxin. Describe and even duplicate the experiments of early scientists such as Charles Darwin, Peter Boysen-Jensen, and F. W. Went.

2. In the fall, leaves from deciduous trees drop off. Auxin plays and important role in *abscission*. As long as leaves and fruits produce auxin, they remain firmly attached to their stems. Auxin appears to be produced in leaves and fruits, and reduction of the hormone causes abscission. Find out what stimuli trigger the reduction of auxin production. What is the importance of leaf drop? How is it a valuable adaptation for deciduous trees? How do the shape and the texture of nondeciduous plant leaves make annual fall leaf drop unnecessary?

TABLE 17.1 FRUIT DEVELOPMENT DATA		
Fertilizer Content	Plants in Group 1 without blossom set	Plants in Group 2 with blossom set
high P, low N		
high P, high N		
low P, high N		
no fertilizer		

17

Notes on
Apical Dominance

Key Facts:

_____
_____
_____
_____

My Results: .

_____
_____
_____
_____
_____
_____
_____

Conclusions:

_____
_____
_____
_____
_____

Results of Try New Approaches:

_____
_____
_____
_____
_____

Notes on Designing My Own Experiment:

_____
_____
_____
_____
_____
_____
_____
_____
_____
_____
_____
_____
_____
_____
_____
_____
_____
_____
_____
_____
_____
_____
_____
_____
_____
_____
_____
_____

Notes on Get the Facts:

_____
_____
_____
_____
_____
_____
_____

18 Geotropism
Plant Movement Due to Gravity

Plants do not move from one location to another, but their leaves, stems, and roots do move. These parts of plants move at such a slow rate that the motion is not noticed. Movement in plants is a response to stimuli, and this movement, or tropism, will be either toward or away from the particular stimulus. Positive tropism is a movement toward stimulus, and negative tropism is a movement away from the stimulus.

In this project, you will learn about geotropism, the response of plants to gravity. You will also determine which parts of a plant exhibit positive and negative geotropism as well as how plants behave in a simulated-gravity environment.

Getting Started

Purpose: To determine how gravity affects plant growth.

Materials

paper towel
12-x-12-inch (30–x-30-cm) sheet of aluminum foil
tap water
mustard seeds
marking pen
1-quart (1-liter) jar

Procedure

1. Fold the paper towel in half twice.

2. Place the folded paper towel in the center of the aluminum foil.

3. Moisten the paper towel with tap water so that it is damp, but not dripping wet.

4. Sprinkle mustard seeds in a line across the center of the damp paper towel (see A in Figure 18.1).

5. Fold the foil around the towel and close up each end of the foil.

6. Use the marking pen to draw an arrow on the outside of the foil package.

7. Stand the package in the jar with its arrow pointing upward (see B in Figure 18.1).

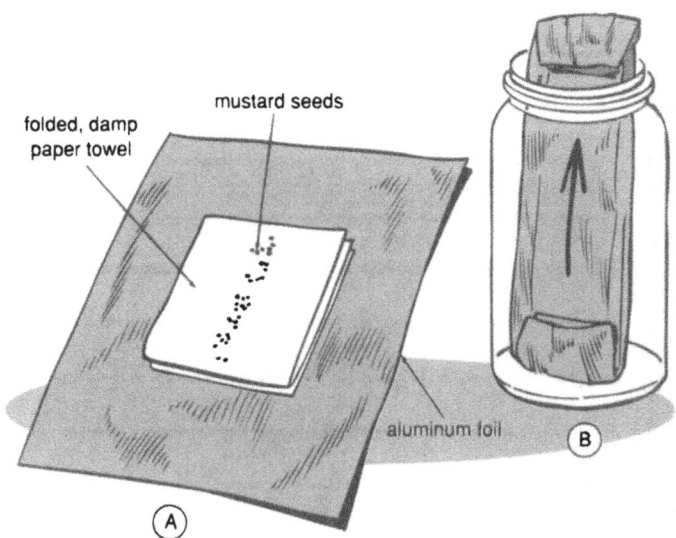

Figure 18.1

8. Place the jar where it can remain undisturbed.

9. After five days, carefully open the package and observe the contents. Make note of the direction of the roots and stems of the seedlings. *Note:* If distinct roots and stems have not yet formed, close the package and reopen it in two days.

Results

The stems of the plants in the package grow up, and the roots grow down.

Why?

Plants are stationary, but they are not motionless. They move slowly, but definitely, in response to stimuli in their environment. This movement in response to a stimulus is called **tropism** and occurs because one part of the plant grows faster than another part. The increased growth is the result of unequal stimulation on opposite sides of the plant.

Auxin is a hormone that causes the cells in a plant to lengthen. Longer cells on one side cause the plant to bend. Different types of cells respond differently to the presence of this hormone. The increase in length of stem cells is directly related to the concentration of auxin in the cells, whereas auxin inhibits rather than increases the growth of root cells.

The growth response of plants to gravity is called **geotropism.** Gravity pulls the growth-stimulating hormone auxin downward toward the lowest part of the stem and root of a plant. More growth occurs in the cells on the lower side of the stem; less growth in the cells on the lower side of the root. The result is that the stem bends one way, upward, and the roots bend in the opposite direction, downward.

Try New Approaches

1a. Can stems and roots change direction? Repeat the experiment preparing four separate foil packages of seeds. After five days, change the position of the packages to further test the effect of gravity on the stems and roots. Support the jars so that the arrows on the packages point up, down, right, and left (see Figure 18.2). Allow the packages to remain undisturbed for another five days. Open the packages and observe the direction of the stems and roots of each plant. Compare the growth of the stems and roots with the control package (the one with its arrow pointing upward). **Science Fair Hint:** Display a data table with diagrams and descriptions of the seedlings. Record the results by referring to the position of the arrow as up (turned 0°), right (90°), down (180°), and left (270°), as in Table 18.1.

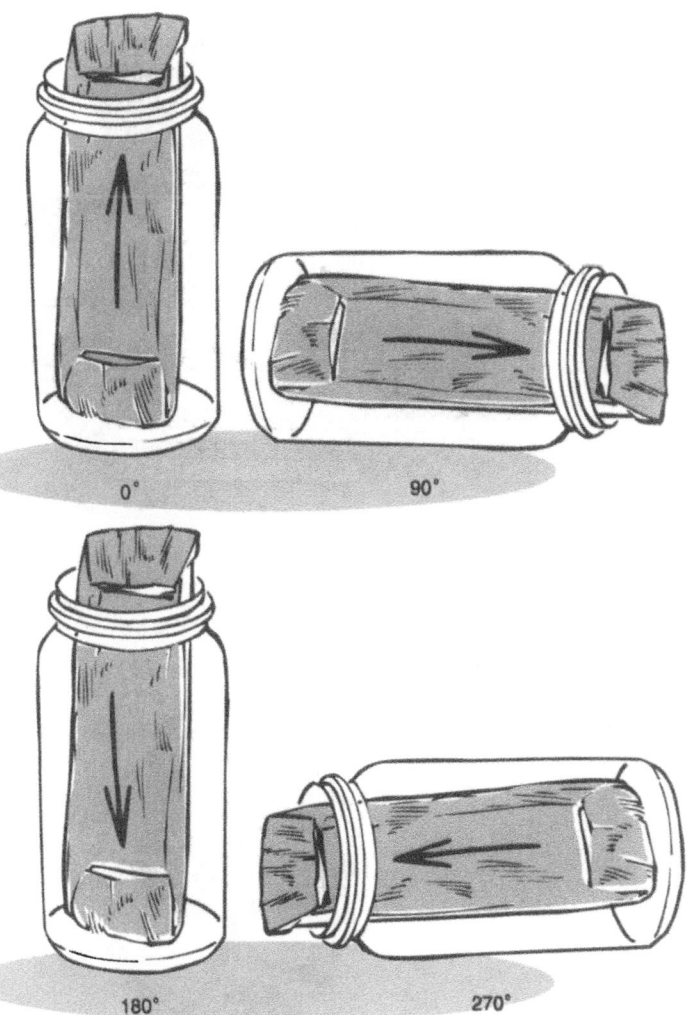

Figure 18.2

TABLE 18.1 GEOTROPISM DATA	
Plant Direction	**Results**
up (0°)	
right (90°)	
down (180°)	
left (270°)	

b. At what rate do the plants move? Repeat the previous experiment using a clear plastic wrap to cover the seeds. Place the packages inside a dark closet or under a cardboard box to prevent the plants from responding to light. Make as many observations as possible each day. Record the day and time for each observation. Use your results to determine whether there is more plant growth at a specific time of day.

2. Does gravity affect one type of plant more than another? Repeat the original experiment using different types of plant seeds, such as radish, pinto beans, or lima beans. *Note:* Different seeds may require more than five days to germinate. **Science Fair Hint:** Use photographs to represent the procedure and results of the experiment.

Design Your Own Experiment

1. Demonstrate the effect of gravity on mature plants by laying one small houseplant on its side (see A in Figure 18.3). Place a covering or netting over the soil of a second small houseplant to keep it in the pot, and then hang the plant upside down (see

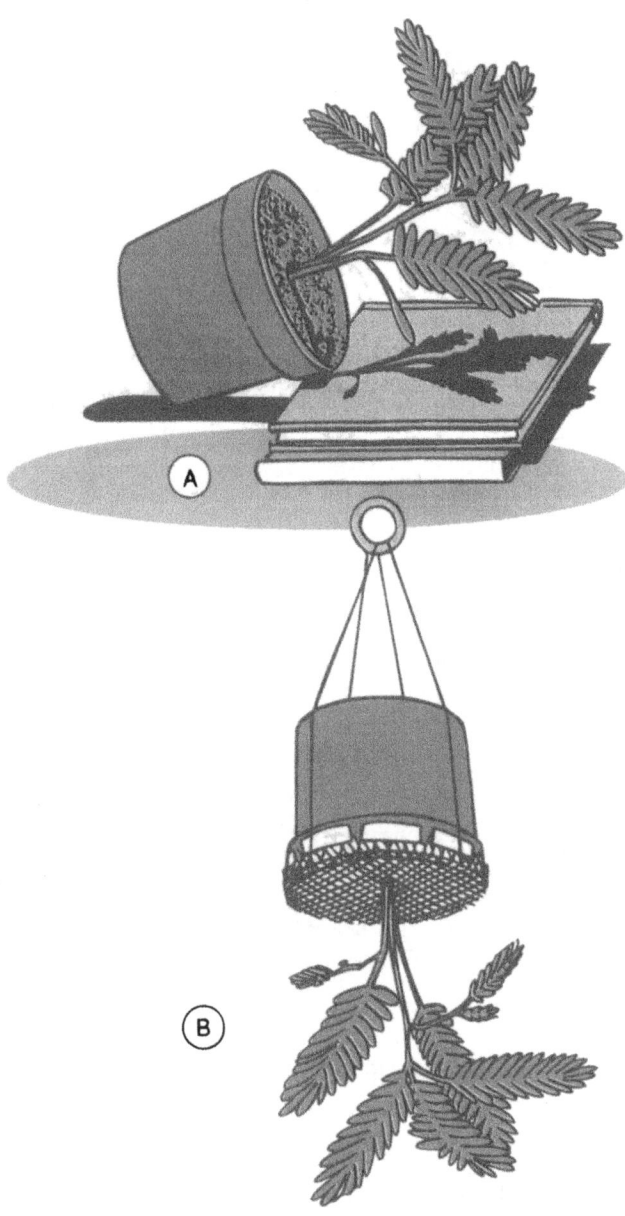

Figure 18.3

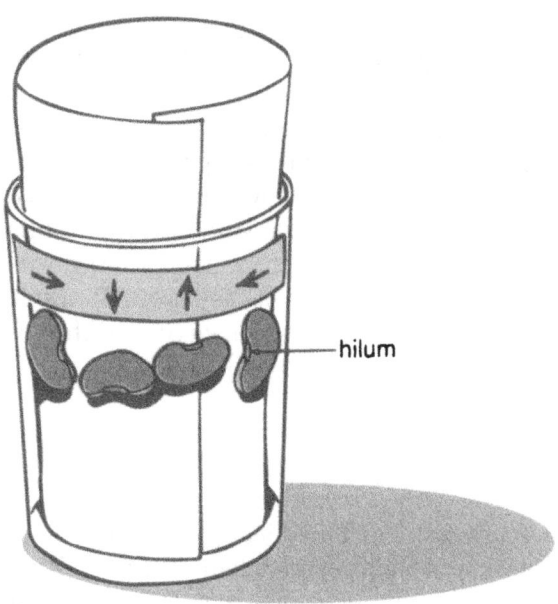

Figure 18.4

paper lining in place. Place a strip of masking tape around the outside of the glass and mark the tape with arrows pointing up, down, left, and right. Place one pinto bean (between the glass and the paper towel) under each arrow. Make sure the bean's hilum (scar on the inside curved side) is pointing in the direction indicated by the arrow (see Figure 18.4) Moisten the paper towel and keep it moist, but not dripping wet. Place the glass inside a dark closet. Observe the direction of the stems and roots daily for seven days.

Get the Facts

1. How does a plant respond to a low-gravity field as in a spacecraft? Write to NASA (L.B.J. Center, 2101 NASA Rd #1, AP-4, Houston, TX 77058) and request information about growing plants in space.

2. How does a simulated-gravity field such as that produced by a rotating space station affect the direction of stem and root growth? A rotating turntable on a record player can produce a simulated field of gravity. More information on growing seeds in this simulated-gravity field can be found in the experiment titled "In or Out?" (p. 50) in Janice VanCleave's *Biology for Every Kid* (New York: Wiley, 1990).

B in Figure 18.3). Place both plants inside a dark closet or under a cardboard box to prevent them from responding to light. Observe the position of the stems of each plant after one week. Carefully remove the soil from around the roots of each plant and observe their direction in relationship to the direction of growth of the stems.

2. Does the direction that a seed is planted affect the growth of the stems and roots? Fold a paper towel and line the inside of a drinking glass with it. Stuff pieces of paper towels into the glass to hold the

Notes on
Geotropism

Key Facts:

_____

_____

_____

_____

My Results:

_____

_____

_____

_____

_____

_____

_____

_____

_____

Conclusions:

_____

_____

_____

_____

_____

Results of Try New Approaches:

_____

_____

_____

_____

_____

Notes on Designing My Own Experiment:

_____
_____
_____
_____
_____
_____
_____
_____
_____
_____
_____
_____
_____
_____
_____
_____
_____
_____
_____
_____
_____
_____
_____
_____

Notes on Get the Facts:

_____
_____
_____
_____
_____

 **Responses of Annelidas**
Segmented Worms

Segmented worms such as the earthworm belong to the phylum Annelida. These worms have a more advanced body structure than most worms. The body of the earthworm is like two tubes, one inside used for digesting food and one outside serving as the body wall. The earthworm has a mouth but no nose, eyes, or ears. Yet it responds to odors and to changes in moisture, temperature, and light.

In this project, you will have the opportunity to unravel these paradoxes by observing and testing on earthworm's responses to various stimuli. You will also test the learning ability of earthworms. *Note:* No earthworms should die from any of the experiments in this project. Handle the worms with care. At the conclusion of the project, place the worms outside in a shady area of soil.

Getting Started

Purpose: To observe an earthworm in a simulated (near-natural) soil environment.

Materials

3 cups (750 ml) of potting soil
1-quart (1-liter) glass jar with lid
1 cup (250 ml) of water
8 to 10 earthworms (from a bait shop, or dig your own)
apple peelings
sheet of newspaper
rubber bands
hammer
large nail

Procedure

1. Pour the soil into the jar.

2. Moisten the soil with water and keep it moist during the entire project.

3. Put the worms into the jar.

4. Place the apple peelings on the surface of the soil and keep the worms supplied with peelings during the entire project.

5. Fold the newspaper so that it fits around the outside of the jar. Secure it with the rubber bands.

6. Use the hammer and nail to make five or six holes in the lid of the jar.

7. Secure the lid and place the jar in a cool place (see Figure 19.1).

8. Remove the paper and observe the jar every day for two weeks.

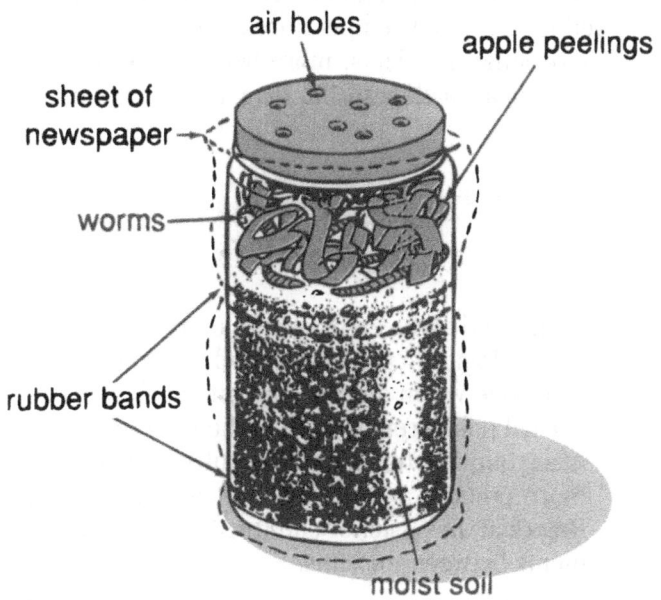

Figure 19.1

Results

The worms start wiggling immediately and burrow into the soil. A network of tunnels can be seen in the soil. The apple peelings disappear, and **casts** (undigested soil deposits) appear on the surface of the soil.

Why?

Earthworms spend their lives in one small area of the ground. In some places there are 50,000 worms per acre of soil. An earthworm's diet consists of the animal and vegetable matter in the soil, which is pulled into the worm's mouth by muscles in its body. Nutrients are extracted as the soil passes through the earthworm's digestive tube and out the other end. In this way, earthworms are very beneficial because of their tunneling and digestive process loosens and aerates the soil.

Try New Approaches

1. How does low temperature affect the activity of the worms? Place the jar of worms in the refrigerator. Make daily observations of the movement of the worms inside the jar for one week. Keep the worms in the refrigerator for the remainder of the project. The body temperature of worms changes with their environment. Their metabolism is reduced in cold temperatures, so they become more lethargic. Thus, more worms can be contained in a smaller area for longer periods of time.

Design Your Own Experiment

1. Which end of the earthworm is more sensitive to odor? Moisten a paper towel and lay it on a table. Place one worm on it. Wet a cotton ball with fingernail polish remover. Hold the wet cotton ball near, but not touching, the anterior end (the more pointed and darker end) of each worm. Repeat at the posterior end (rear end) and at segments between the ends.

2. Does the earthworm prefer wet or dry surfaces? Lay two pieces of wet and dry paper towels next to each other, but not touching, on a table. Place an earthworm across the two surfaces, anterior end on the wet towel, and observe the response. Then reverse the worm's position so that the anterior end is on the dry towel. Again, observe the response.

3a. Does the earthworm respond faster at a higher or lower body temperature? Prepare two containers of worms by placing soil and three worms in two separate baby food jars. Place one jar in a refrigerator and keep the other at room temperature. After 24 hours, use the worms from each jar to test their response to odor and preference for wet or dry paper.

b. Use a stereomicroscope (dissecting microscope) to measure the change in heart rate of the worms at different body temperatures. Place one worm at a time under the microscope. Find the earthworm's dorsal aorta (the blood vessel running along the uppermost region of the back). You should see a wavelike contraction moving from the posterior end to the anterior end. Each wave is a single "heartbeat." Count the beats in one minute. Measure the temperature of the soil. Make additional temperature changes by sitting the containers in a bowl of ice water. Do not heat the water because you will injure the worms. Prepare a graph using temperature as the independent variable (x-axis) and heartbeat as the dependent variable (y-axis).

4. Can an earthworm learn simple, consistent choices when confronted with alternatives? Use shoe boxes to construct a T-maze (see Figure 19.2). Cover the top with several layers of red cellophane and restrict direct light on the maze. Use a "D" cell battery and wire to supply a brief electric shock when the worm touches the wire. Earthworms can learn to take the arm of the maze leading to the reward (darkness and moisture) and away from punishment (light, dryness, sandpaper, and electric shock). The development of this response is slow and will require many trials. Separate the worms that are learning the maze and place them in their own container. It is thought that earthworms can learn to make the correct choice 90% of the time. How well do your worms learn? How long do they retain what they learn? Display the maze and photographs showing the worms during a typical learning lesson as well as a graph comparing the trials for each worm in training.

Get the Facts

1. The American earthworm belongs to the genus *Lumbricus*. Charles Darwin first showed that this worm is important because it aerates the soil by digging tunnels and aids in the growth of plants by dragging seeds from the surface into the damp soil where they can germinate. Use an encyclopedia to find out more about the benefits of earthworms. How deep do they tunnel? How much soil do they bring to the surface annually?

2. Earthworms have a nervous system but no obvious sense organs such as eyes. But the worms

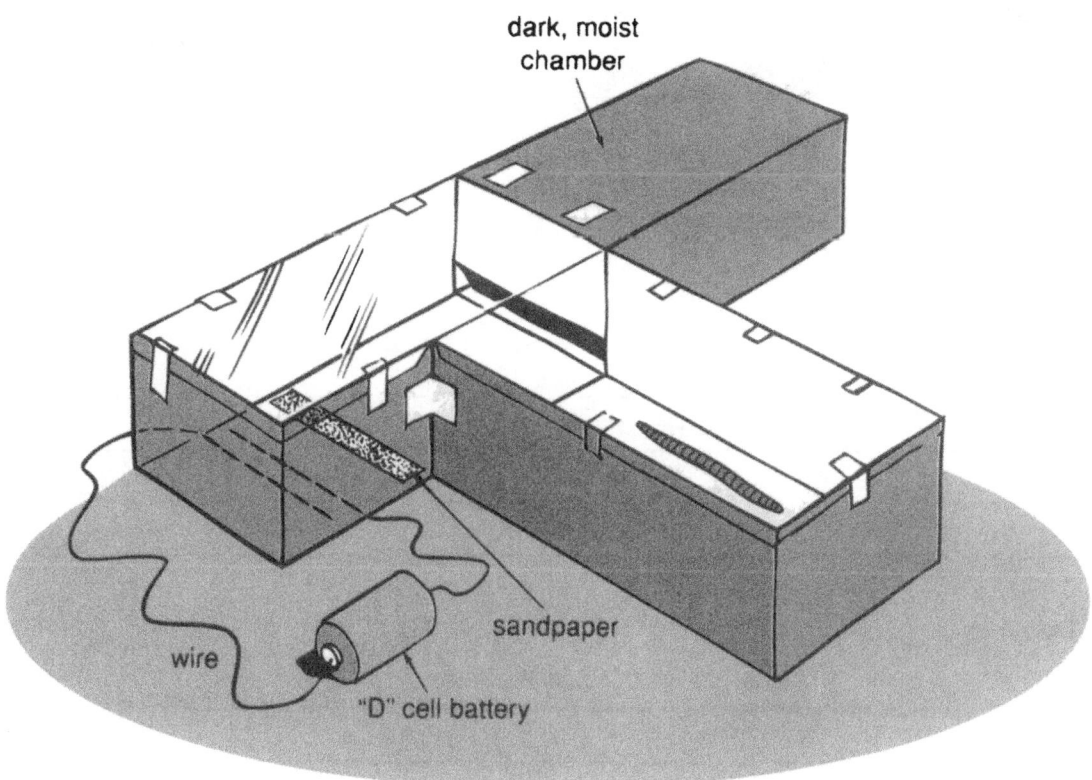

dark, moist
chamber

wire

"D" cell battery

sandpaper

Figure 19.2

respond to light stimuli? How? Use a biology text to find information about the sensory responses of worms. Also read the experiment title "Night Crawlers" (pp. 124–125) in Janice VanCleave's *Biology for Every Kid* (New York: Wiley, 1990).

Notes on
Responses of Annelidas

Key Facts:

_____

_____

_____

_____

My Results:

_____

_____

_____

_____

_____

_____

_____

_____

Conclusions:

_____

_____

_____

_____

_____

Results of Try New Approaches:

_____

_____

_____

_____

_____

Notes on Designing My Own Experiment:

_____
_____
_____
_____
_____
_____
_____
_____
_____
_____
_____
_____
_____
_____
_____
_____
_____
_____
_____
_____
_____
_____
_____
_____
_____
_____

Notes on Get the Facts:

_____
_____
_____
_____
_____
_____

20 Aquatic Respiration
Breathing Rate of Goldfish

Aquatic (pertaining to water) bony fish extract oxygen from water as terrestrial animal lungs draw oxygen from air. But gills are more efficient at extracting oxygen from water than lungs are at getting oxygen from the air. Gills extract about 80 % of the oxygen in water they take in, while human lungs extract only about 20 % of the oxygen from air they breathe.

In this project, you will determine the breathing rate of a goldfish. You will determine the effect of variables on the rate of breathing, such as fish size and type, water temperature, and the fish's excitement level.

Getting Started

Purpose: To determine the breathing rate of a fish by observing the opening and closing of its mouth and operculum.

Materials

medium-size goldfish in a small fishbowl
timer

Procedure

Note: The goldfish should be in a small fishbowl that has been properly prepared according to instructions from a professional at a pet store or from a fish manual. Feeding the fish should be considered part of the normal maintenance of the fish and not as part of a project experiment.

1. Locate the fish's mouth (see Figure 20.1).

Figure 20.1

2. Count the number of times the fish opens and closes its mouth over a period of 2 minutes. Record the number of openings in a Fish Breathing Rate Data table, like the one in Table 20.1.

TABLE 20.1 FISH BREATHING RATE DATA					
	Mouth openings in 2 minutes				
	Test 1	Test 2	Test 3	Test 4	Average
Medium Fish					

3. Repeat step 2 three times, then average the number of mouth openings.

4. Calculate the breathing rate of the fish per minute by taking the average rate for two minutes and dividing it by 2. For example, if the average is 210 mouth openings in 2 minutes, the breathing rate per minute would be:

$$\text{breathing rate} = 210 \text{ openings} \div 2 \text{ minutes}$$
$$= 105 \text{ openings /minute}$$

5. Repeat steps 1 through 4 counting the opening and closing of the fish's **operculum** (protective flap of skin over a fish's gills).

Results

The fish's mouth and operculum open and close an equal number of times. The breathing rate in the example is 105 openings per minute.

Why?

External respiration is a process used by organisms to exchange gases with the environment. **Breathing** (mechanical process of moving gases into and out of the body) is an essential part of the external respiration in most animals. The respiratory system of some **aquatic** (pertaining to water) organisms, including bony fish, such as a goldfish, consists of a set of gills beneath a protective flap called the **operculum. Gills**

are **organs** (structures in organisms that perform a function) through which fish breathe. One set is on either side of the fish's body. The gills contain many **capillaries** (tiniest blood vessels). As water flows over the gills, oxygen from the water enters the blood by passing through the walls of these capillaries and the waste gas, carbon dioxide, in the blood leaves the capillaries and enters the water. Thus an exchange of gases occurs in the gills.

A fish opens its mouth to inhale water, not to drink it. "Yawning" movements draw water into the mouth. When the mouth is open the operculum is closed. Then the mouth closes and the operculum opens. The water that came in through the mouth then passes over the gills and out the operculum. The breathing rate, or the number of times the mouth or the operculum open and close per minute, was determined for the fish in this experiment.

Try New Approaches

1. How does the size of a fish affect its breathing rate? Determine this by repeating the experiment using goldfish of different sizes; use fish that are larger and smaller than the one in the experiment. Let the medium-size fish be the control and compare the breathing rates of the other fish with that of the control fish.

2. Is the breathing rate of different kinds of fish the same? Determine this by repeating the original experiment using different kinds of fish. Select fish that are of comparable size to the medium-size goldfish in the original experiment.

Design Your Own Experiment

1a. During the winter, a fish's water environment gets colder. Design an experiment to determine if a change in water temperature affects the respiration rate of fish. One way is to prepare a testing tank by filling a 1-quart (1-liter) wide-mouth jar about three-fourths full with distilled water. Allow the water to sit for 30 minutes or more until it is at room temperature. Using a fishnet, carefully transfer the medium-size fish from the original experiment into the test jar. Place a clean thermometer into the jar and set the jar in a bowl

half filled with water. Slowly lower the temperature of the water inside the testing jar by adding a few small pieces of ice to the bowl (see Figure 20.2). When the ice pieces melt, use the thermometer to gently stir the water inside the testing jar so that the water's temperature is the same throughout. Allow the testing jar to remain undisturbed for 5 minutes to allow the fish to calm. Then, determine the respiration rate of the fish at this temperature. Continue cooling the jar by small degree intervals until the temperature is about 41°F (5°C). Do not lower the temperature to freezing at 32°F (0°C). After the experiment is complete, allow the testing jar to sit so that it returns to room temperature before returning the fish to its regular container. Prepare and display a graph showing the average breathing rate of the fish at each temperature (see Figure 20.3 on the following page).

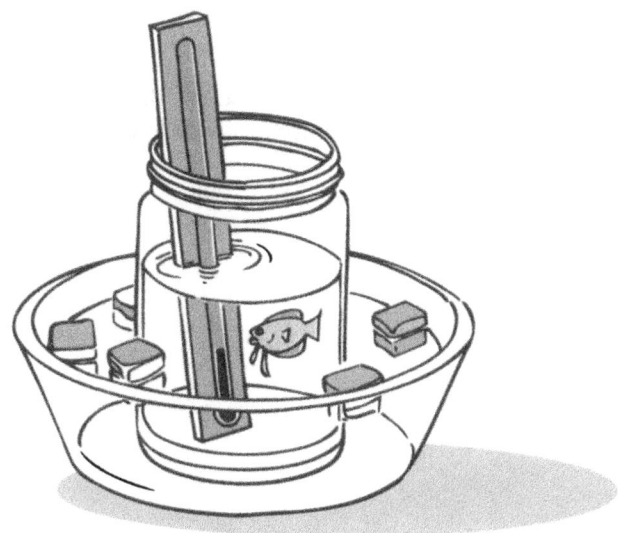

Figure 20.2

b. Repeat the previous experiment to determine the effect of temperature on different sizes of fish as well as different kinds of fish. On a line graph, use a different color line for each different fish tested.

2a. Animals breathe faster when they exercise as well as when they get excited. Design an experiment to determine if being excited affects the breathing rate of a fish. One method is to cover the fishbowl of the medium-size fish with a dark cloth and restrict as much as possible noises and movement outside the bowl. After five or more

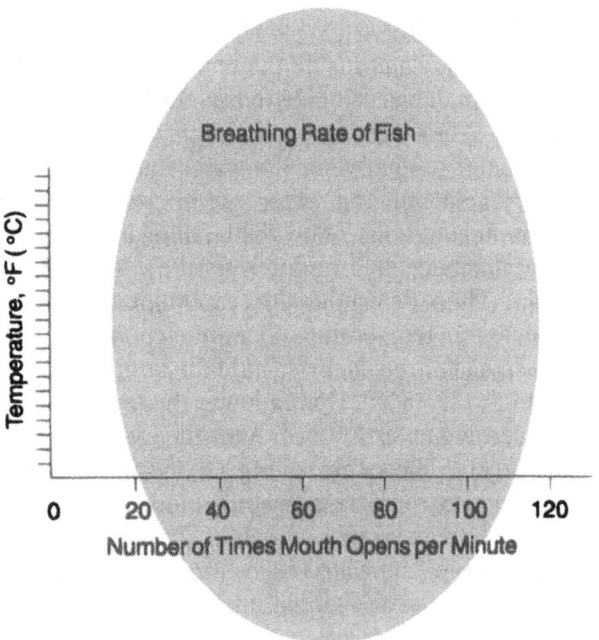

Breathing Rate of Fish

Temperature, °F (°C)

0   20   40   60   80   100   120

Number of Times Mouth Opens per Minute

Figure 20.3

minutes, quietly position yourself so that you can immediately begin counting mouth movements as you quickly remove the cloth. Calculate the breathing rate per minute of the startled fish.

b. Repeat the previous experiment using fish of different sizes as well as different kinds of fish.

Get the Facts

Gas exchange in an organism supports cellular respiration by supplying oxygen and removing carbon dioxide. The source of oxygen, called the *respiratory medium,* is water for fish and air for terrestrial animals. The part of an animal's anatomy where gas exchange with the respiratory medium occurs is called the *respiratory surface,* which is the gills in a fish. The greater the respiratory surface, the greater is the gas exchange. What is the purpose of gill rakers and how do they affect gas exchange? What does ventilation of gills mean? Why is it necessary to keep the respiratory surface wet? What is countercurrent exchange? For more information, see a biology text.

Notes on
Aquatic Respiration

Key Facts:

_____

_____

_____

_____

My Results:

_____

_____

_____

_____

_____

_____

_____

_____

Conclusions:

_____

_____

_____

_____

_____

Results of Try New Approaches:

_____

_____

_____

_____

Notes on Designing My Own Experiment:

_____
_____
_____
_____
_____
_____
_____
_____
_____
_____
_____
_____
_____
_____
_____
_____
_____
_____
_____
_____
_____
_____
_____
_____
_____
_____

Notes on Get the Facts:

_____
_____
_____
_____
_____
_____

Notes on Designing My Own Experiment:

_____

_____

_____

_____

_____

_____

_____

_____

_____

_____

_____

_____

_____

_____

_____

_____

_____

_____

_____

_____

_____

_____

Notes on Get the Facts:

_____

_____

_____

_____

_____

22 Ecosystem Interactions

In nature, there is a constant interaction among animals, plants, fungi, protists—microorganisms not easily classified as fungi, plant or animal—and their environment. A specific area where living and nonliving things interact is called an ecosystem. In this ecosystem, there is a direct relationship between the community of living organisms and its surroundings. The organisms affect their physical surroundings just as they are affected by them.

In this project you will observe some of these interactions in an open ecosystem and investigate populations within a sector. You will also determine populations of organisms by using the technique of random sampling.

Getting Started

Purpose: To select and lay out a study area.

Materials

compass
measuring tape
hammer

12 wooden or metal tent stakes
300 yards (300 m) of cord

Procedure

1. Select a study area that has a variety of plant life. (A wooded area was chosen by the author, but any ecosystem can be used.)

2. Use the compass to determine which direction is north.

3. With the measuring tape, measure an area 30 × 30 yards (30 × 30 m). The plot of ground should be laid out so that it is aligned in a north-to-south direction.

4. At each of the four corners of the plot, hammer one stake into the ground, leaving about 6 inches (15 cm) of the stake aboveground.

5. Attach the end of the cord to one stake, loop it around the other stakes, and tie it to the starting stake to enclose the plot.

6. Use the measuring tape to divide the sides of the plot into 10-yard (10-m) sections.

7. Drive one stake into the ground at each 10-yard (10-m) interval along all sides of the plot (see Figure 22.1).

8. Using cord to join opposite stakes, divide the plot into nine equal subplots.

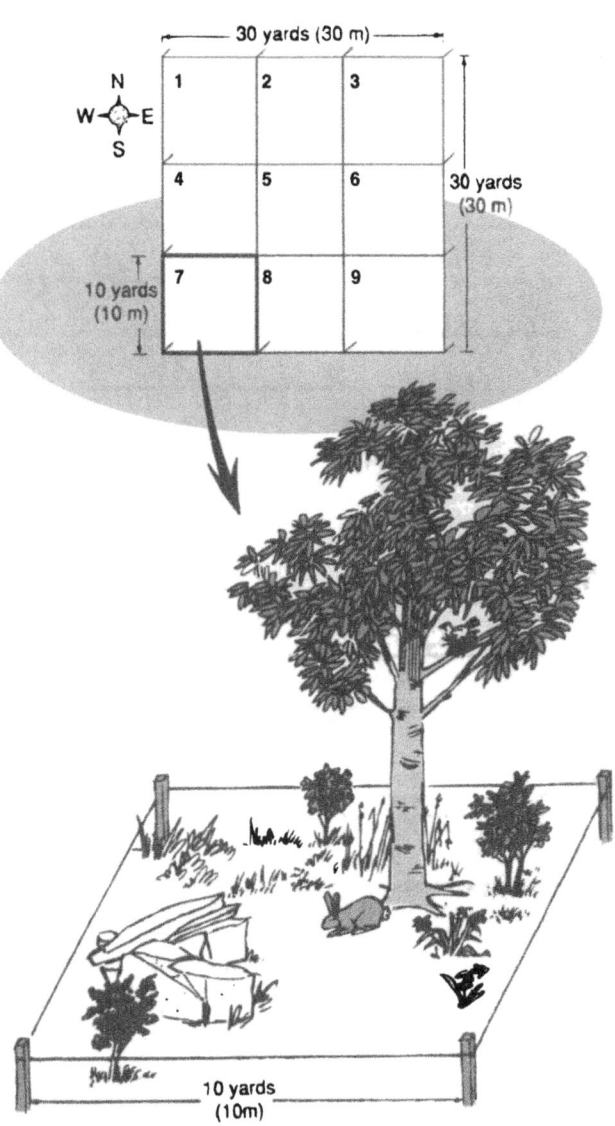

Figure 22.1

Results

A sampling plot of ground is selected, measured, and subdivided.

Why?

An **environment** is all the external factors affecting an organism. These factors may be **biotic factors** (living organisms) or **abiotic factors** (nonliving things including water, soil, climate, light, and air). All organisms living in a particular environment, such as a forest or desert, together with the abiotic factors that affect them, are collectively known as an **ecosystem.** In an ecosystem, energy and nutrients flow between the abiotic and biotic environment.

A **sampling plot** is a select area of an ecosystem that allows you to study the biotic and abiotic factors in it. During a field study, measurements of physical factors and organisms are taken. Separate information taken from each subplot, when studied as a whole, provides a clear picture of the abiotic and biotic factors within the plot and gives clues to the greater ecosystem surrounding the plot.

Try New Approaches

On graph paper, sketch the plot. Identify each subplot with a number. Indicate the compass directions on the sketch with arrows. Make separate sketches of each subplot. Note prominent land features such as trails, open areas, erosion, and streams.

Design Your Own Experiment

1. Measure the angle of the **slope** (steepness of an inclined surface) of each subplot. Do this by constructing and using an Abney level. Make the level by using masking tape to attach a protractor to the center of a yardstick (meterstick). Tie a 12-inch (30-cm) piece of string to the center of the protractor and attach a washer to the free end of the string (see Figure 22.2). Hold the measuring stick parallel to the ground of each subplot while a helper reads the angle on the protractor made by the hanging string. When the slope is zero, the string hangs straight down across the 90° mark. To determine the angle of slopes, subtract

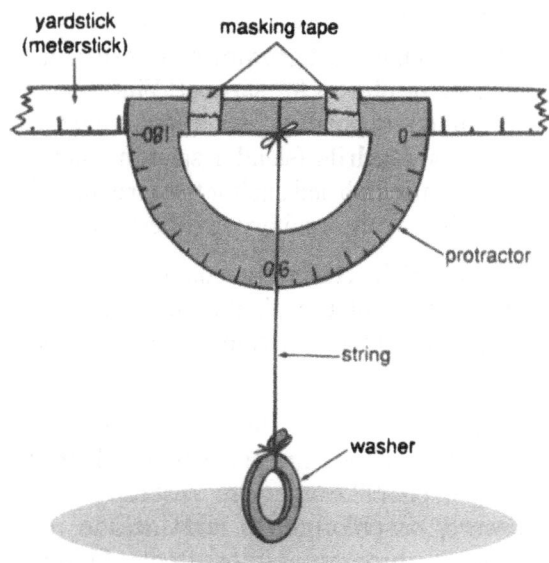

Figure 22.2

the angle reading from 90°. For example, if the protractor reads 50°, the angle of the slope is 90° – 50°, or 40°.

2a. Observe the types of plants in each subplot. Collect leaf samples from the plants and, using plant field guides, identify the types of each plant present. Land plants are commonly classed as trees, shrubs, herbs, or vines according to the size and type of stem. Land plants are either a **woody plant** (a plant containing stems with a large amount of wood) with relatively hard stems or a **nonwoody plant** (a plant containing stems with a small amount of wood) with relatively soft stems. **Wood** is the part of a plant made of xylem tubes (tubes that transport water and other nutrients from the root throughout a plant and provide support). A description of the four types of land plants follows:

- **Trees**—Plants with a single tall, woody stem called a trunk that grows from the ground and is able to stand erect without support. Other woody stems branch from the trunk.

- **Shrubs**—Short, woody plants usually with several stems branched near the ground that are able to stand erect without support. Also called bushes.

- **Herbs** or **herbaceous plants**—Names for nonwoody plants. The stems of these plants remain relatively soft and include grasses and flowers, such as daisies and dandelions.

- **Vines**—Flexible, weak-stemmed plants whose long, slender, fast-growing shoots rely on other plants or objects for support. Vines may wind around supporting structures, and/or use thorns, **tendrils** (slender spiraling stems), or hooks to climb and anchor to a structure. Vines can be woody or nonwoody (herbaceous).

Science Fair Hint: You could display labeled photographs of the plants from each subplot along with preserved samples of the leaves or leaf rubbings.

b. Observe and record the presence of different fungi and lichens in each subplot. **Fungi** is a group of plantlike organisms that have no leaves, flowers, or chlorophyll, and include mildew, mold, mushrooms, and toadstools. **Lichen** is a plantlike organism made of fungus and **algae** (a plantlike organism with no leaves, stems, or roots, but that has chlorophyll necessary for photosynthesis). See a biology text for more information about fungi and lichen.

3a. Record the actual presence or indications of the presence of animals. Note the presence of tracks, holes, sounds, claw marks, hair, nests, and so on. Cautiously turn over rocks and logs to discover insects, reptiles, and amphibians. Remember that you are an observer and do not want to change this ecosystem, so return rocks and logs to their original positions.

b. Make molds of tracks by pouring plaster of paris into the tracks. Mix this liquid by following the directions on the box. You could use the molds as part of a project display.

Get the Facts

1. When an organism feeds upon another living thing, there is a transfer of materials and energy. This transfer from organism to organism is called a *food chain*. Find out more about the food chain. What is a food web? What is a pyramid of energy and a pyramid of numbers? Use the information collected in your plot to construct a pyramid of numbers. For information, see *Janice VanCleave's Ecology for Every Kid* (New York: Wiley, 1996), pp. 37–46.

2. Energy is constantly lost in an ecosystem and has to be replaced by sunlight. There is a fixed amount of matter within the system. Find out how matter is recycled in an ecosystem. For information, see a biology text.

Notes on
Ecosystem

Key Facts:

_____
_____
_____
_____

My Results:

_____
_____
_____
_____
_____
_____
_____
_____

Conclusions:

_____
_____
_____
_____
_____

Results of Try New Approaches:

_____
_____
_____
_____
_____

Notes on Designing My Own Experiment:

_____
_____
_____
_____
_____
_____
_____
_____
_____
_____
_____
_____
_____
_____
_____
_____
_____
_____
_____

Notes on Get the Facts:

_____
_____
_____
_____
_____
_____

23 Preservatives
Food Additives

In our complex world, food products have to be shipped for long distances and/or stored for periods of time. It would be difficult to transport and store most kinds of food without using preservative additives.

In this project, you will have the opportunity to test the effectiveness of calcium propionate, a food additive that inhibits mold growth in bread. The effect of this additive in different types of bread as well as the effect of temperature on the preservative will be determined. You will also analyze other methods of preserving food.

Getting Started

Purpose: To determine how effective the food additive calcium propionate is in inhibiting the molding of bread.

Materials

paper towels
water
6 plastic 1-gallon (4-liter) size resealable bags
6 slices of white bread with calcium propionate
6 slices of white bread without calcium propionate or any other preservatives
marking pen
masking tape
magnifying lens (hand-held type)

Procedure

CAUTION: *Do not do this project if you are allergic to mold.*

1. Moisten one paper towel with water and lay it inside one plastic bag.

2. Place one slice of bread with calcium propionate and one without side by side on top of the moistened paper towel inside the bag.

3. Zip the plastic bag closed.

4. Use the marking pen and masking tape to label the slices of bread with and without calcium propionate. Place the label on the outside of the bag above the indicated slice of bread.

5. Repeat the procedure (steps 1 through 4) preparing five additional bags with two slices of bread, one slice of bread with calcium propionate and one without, inside each bag. *Note:* Six bags is not a significant number, but it does provide enough samples to verify your results.

6. Keep the six bags of bread closed and at room temperature.

7. Examine the slices each day with the magnifying lens.

8. Continue observing the bread for two weeks or until every slice has become moldy. Record the length of time required for each slice to mold. Discard the unopened bags.

Results

Given enough time, all of the bread slices become moldy. However, the slices with calcium propionate mold more slowly (see Figure 23.1).

Figure 23.1

137

Why?

Food additives are natural and synthetic chemicals added to food to supply nutrients, to enhance color, flavor, or texture, and/or to prevent or delay spoilage. **Spoilage** is the ruining of food as a result of chemical changes often owing to the presence of organisms such as fungi and microbes. A **food preservative** is a food additive, such as calcium propionate, used to maintain freshness and extend the shelf life of package food. Organisms that spoil food are fussy about their diet, and different **species** (group of similar organisms) can be found on specific foods.

Calcium propionate is one of the food additives on the U. S. Food and Drug Administration's **GRAS** ("Generally Recognized As Safe") **list.** At low concentrations, it is considered harmless to humans but inhibits the reproduction and growth of mold. The addition of calcium propionate to bread allows the product to be stored for longer periods of time.

Calcium propionate is a preferred preservative for bread because it retards the rapid growth of bread mold, increases the content of calcium, and avoids the possibility of decreasing gas formation during baking, thus allowing the bread to rise normally.

Try New Approaches

1. How effective is calcium propionate in breads other than white bread? Repeat the experiment using different types of bread. such as wheat, rye, and potato. **Science Fair Hint:** Record results of the original as well as this experiment and display the data tables.

2. Does temperature affect the effectiveness of the preservative calcium propionate? Repeat the original experiment using two sets of bread samples. Place one set in the refrigerator and place the second set in a warm area such as on top of a refrigerator. **Science Fair Hint:** Take photographs as each experiment progresses to show the changes in the food as a result of containing or not containing calcium propionate.

Design Your Own Experiment

Note: Avoid touching the food test samples. The food samples are not edible and should be discarded after the experiments.

1. Drying foods was one of the earliest methods of food preservation. This method is based on the fact that the organisms known to spoil food do not grow and reproduce without water. Find out more about procedures for preserving food by drying. Place dried and fresh food samples in open dishes (beef jerky and strips of fresh beef make good samples). Examine the food samples daily for evidence of spoilage and determine the effectiveness of drying as a method of preserving.

2a. Table salt (sodium chloride) and vinegar (acetic acid) are used as preservatives. Test the effectiveness of these preservatives on inhibiting microbe growth. Dissolve one chicken bouillon cube in 1 cup of (250 ml) of hot water. Divide the solution equally between three clear glasses. Add 1 teaspoon (5 ml) of salt to one glass and 1 teaspoon (5 ml) of vinegar to the second glass. The third glass is the control. Label the glasses accordingly. Place the glasses in a warm place and examine their contents daily. Spoilage due to the presence of microbes results in a solution that looks cloudy, has an odor, and often contains gas bubbles.

b. Does the amount of preservative added change the results? Repeat the experiment two times, first using one half as much salt and vinegar, and then using twice as much salt and vinegar.

3. Many foods are spoiled by the growth of various fungi in the food. Favorable conditions such as moisture and warm temperature encourage rapid reproduction of fungi. Sugar is used as a preservative for fruits because it aids in removing moisture from the cells of the fruit. Fungi are less likely to grow in the dryer fruit. Demonstrate sugar's ability to dehydrate fruit cells by peeling an apple and cutting it into small pieces. Place the pieces into a jar and add ¼ cup (62 ml) of granulated sugar and stir. Secure the lid on the jar. After about 24 hours, the apple pieces will be surrounded by a thick solution of the sugar dissolved in water from the fruit's cells. Use diagrams showing the changes in the jar to represent the dehydrating property of sugar. As part of a display, show pictures of products that use sugar as a preservative. *Note:* Salt also dehydrates and, like sugar, preserves by drying out food such as meat.

Get the Facts

1. The U.S. Food and Drug Administration (FDA) regulates the use of food additives. Nitrates and nitrites are used as preservatives for meat; they also give meat a healthy-looking red color. These preservatives are potentially dangerous because they can lead to the production of cancer-causing chemicals in the digestive system. FDA regulations allow foods to contain up to 500 parts per million (ppm) of nitrate and 200 ppm of nitrite. It is considered safe for adults to consume these preservatives at this low level, but foods for babies under one year of age should contain no nitrate or nitrite additives. Find out more about nitrates and nitrites. Could other safer methods of preserving meats be used?

2. The food additive sodium benzoate is on the GRAS list. This chemical occurs naturally (in very small amounts) in many foods, such as prunes and cranberries. It is naturally present and part of the metabolic process in the human body. Find out more about sodium benzoate. What conditions must exist for it to be effective? List foods it is used in. What is considered a harmless level for human consumption?

3. The FDA allows preservatives to be added to foods at specific levels. How are these safe levels of consumption determined?

4. Five types of microorganisms that cause food to spoil are staphylococci, bacilli, molds, streptococci, and yeasts. These organisms feed on much the same substances as do humans and, when present, can bring about undesirable chemical and physical changes in food. With the right conditions, they can double in number every 20 or 30 minutes. Find out more about these food spoilers and how they can be controlled. You could prepare a data table for display showing diagrams of the five types of microorganisms and listing specific types of food spoilage, such as botulism (one of the most serious types of food poisoning) or simple fermentation by yeast. (See the sample data table in Figure 23.2.)

5. Find out more about advancements in preserving foods. Compare home food preservation methods in the past and at present. Some things have changed very little, such as the smoking of meats on an open fire. You could use diagrams to compare old and new techniques.

6. In the late eighteenth century, Napoleon offered 12,000 francs for a new way of preserving food. Nicolas Appert, a Parisian chef, won the money with his heat-preserved foods. Find out how this man without any knowledge of microbes and their role in food spoilage was able to preserve food.

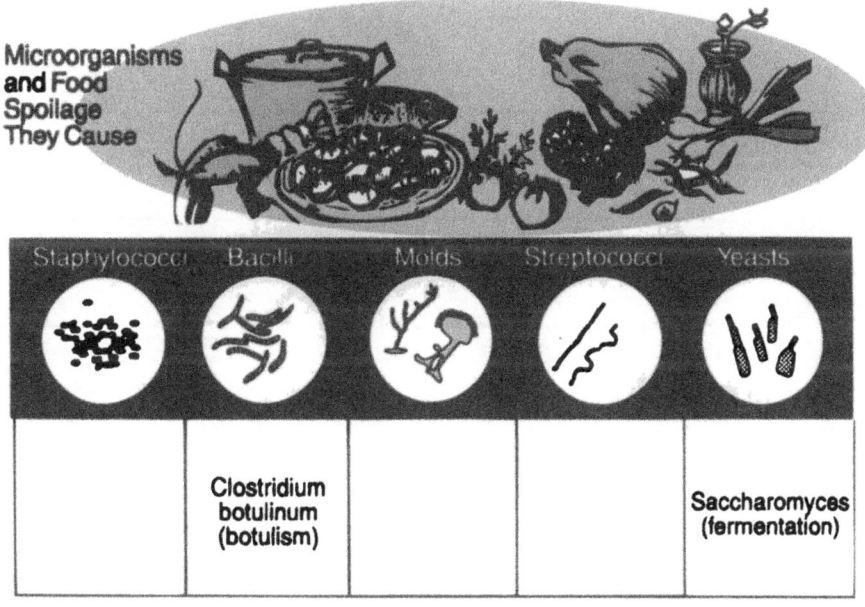

Figure 23.2

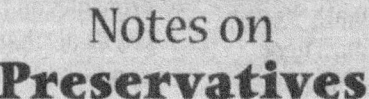

Notes on
Preservatives

Key Facts:

_____

_____

_____

_____

My Results:

_____

_____

_____

_____

_____

_____

_____

_____

Conclusions:

_____

_____

_____

_____

_____

Results of Try New Approaches:

_____

_____

_____

_____

_____

Notes on Designing My Own Experiment:

_____
_____
_____
_____
_____
_____
_____
_____
_____
_____
_____
_____
_____
_____
_____
_____
_____
_____
_____
_____
_____
_____

Notes on Get the Facts:

_____
_____
_____
_____
_____
_____
_____

Chemistry

24 Phase Changes
Effects of Solutes

Phase changes, such as melting, freezing, evaporation, and condensation, occur whenever the physical phase (gas, liquid, or solid) of a substance changes. Solutes may play a significant role in these changes. Salt, for example, is used to melt ice on sidewalks and to make ice colder in an ice-cream maker.

In this project, you will determine the freezing and boiling points of water and the effect of salt and other solutes on these temperatures. Colligative properties, which depend only on the number of particles in a solution, will be studied and used to calculate the freezing-point depression and boiling-point elevation of solutions.

Getting Started

Purpose: To determine the boiling point of water and plot a time-temperature graph of the phase change.

Materials

1 quart (1 liter) of distilled water	candy thermometer
2-quart (2-liter) cooking pot	stove
fondue fork (or heavy wire)	

Procedure

1. Pour the distilled water into the pot.

2. Place the fondue fork (or heavy wire) across the pot and clip the thermometer to the fork so that the bulb is suspended in the middle of the water (see Figure 24.1).

3. Read and record the temperature of the water in a data table such as Table 24.1.

4. Place the pot of water on the stove.

5. Heat at a medium temperature.

6. Every 15 seconds, read and record the temperature until the water stops getting hotter and three readings of the same temperature are recorded. *Note:* This should take about six minutes.

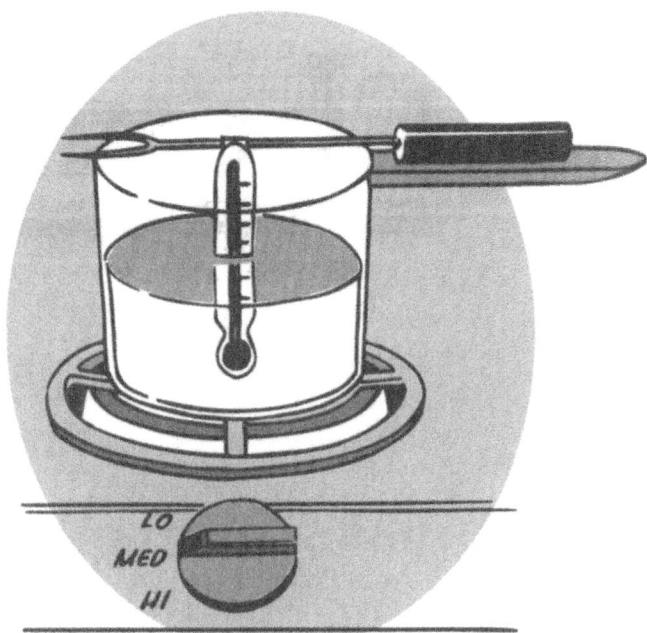

Figure 24.1

TABLE 24.1 DATA TABLE		
Time (seconds)	Temperature (°F or °C)	Observation
0		
15		

7. Observe and make note of the appearance of the water each time the temperature is recorded

8. Plot the results on a graph with the temperature on the vertical axis and the time on the horizontal axis (see Figure 24.2).

Results

As the temperature increases, tiny bubbles appear on the bottom and around the sides of the pot. The bubbles increase in size and begin to break at the surface of the water as the water temperature reaches (exactly or approximately) 212°F (100°C). The bubbles continue to break at the surface, but the temperature remains constant.

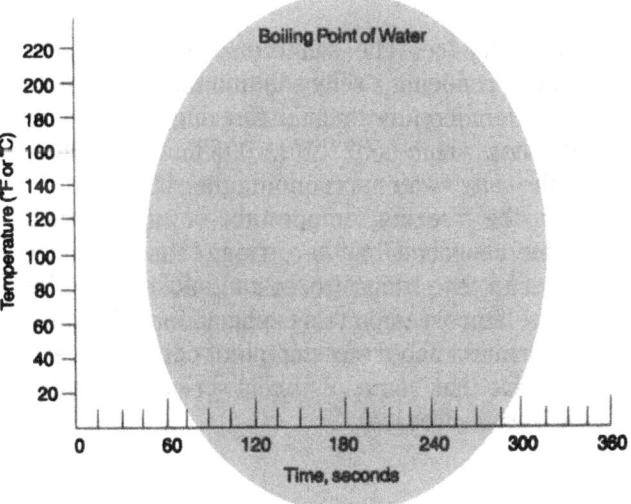

Figure 24.2

Why?

Certain conditions are required to effect a **phase change** of matter from on **physical phase** (gas, liquid, or solid) to another. Heating the water in the pot causes vapor bubbles to form on the bottom and around the sides of the pot where the water is the hottest. As the bubbles rise through the cooler water, they not only cool but also are pressed from all sides by the water molecules, causing them to collapse (see Figure 24.3).

As the water temperature increases, the pressure inside the vapor bubbles increases. Eventually, the **boiling point** is reached (a temperature at which the pressure inside the bubbles equals the atmospheric pressure outside). At the boiling point, the bubbles do not collapse but escape from the surface of the water.

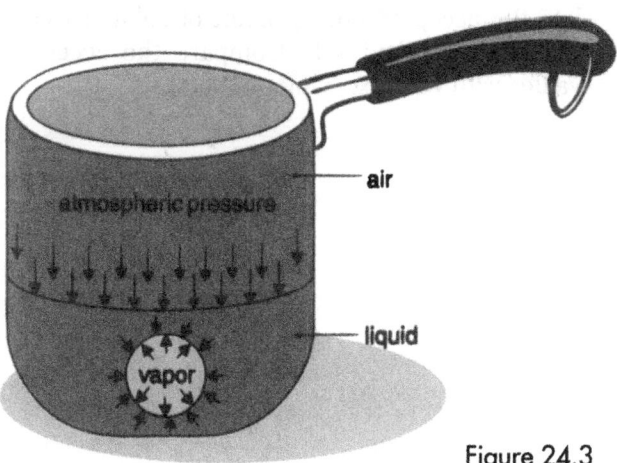

Figure 24.3

During boiling, the temperature of the vapor is the same as that of the liquid. The **kinetic energy** (energy in motion) of liquids is less than that of vapor. Because of the lesser amount of kinetic energy, liquid molecules move less and are more closely bonded together. It takes extra energy to physically break the bonding between liquid molecules and change them into gas. At the boiling point of a liquid, the energy applied by the heating source does not increase the water's temperature but is used in the phase change from liquid to gas.

Try New Approaches

1. Does increasing the heat source raise the temperature of the water? Repeat the experiment using a higher heat setting on the stove.

2a. Does adding a **solute** (substance dissolved in a solution) to the water affect the boiling-point temperature of the water? Repeat the original experiment adding ½ cup (125 ml) of sodium chloride (table salt) to the water.

b. Does using a different solute change the results? Repeat the original experiment adding ½ cup (125 ml) of sucrose (table sugar) to the water. **Science Fair Hint:** Use neatly labeled graphs as part of a display.

Design Your Own Experiment

1a. At what temperature does the **freezing** (the physical change from a liquid to a solid) of water occur? The temperature of a mixture of ice and water will change until the freezing point (or the melting point) is reached. At the freezing point, the temperature of the solution remains constant as energy is used in the phase change. Demonstrate this by filling an ice-cube tray with distrlled water and placing it in a freezer overnight. Fill a drinking glass half full with these pure ice cubes (made with distilled water) and cover the ice with distilled water. Insert a thermometer into the glass of ice water. Set the glass in a large can. Fill the can with ice (this does not have to be pure ice) mixed with 1 cup (250 ml) of table salt.

Gently stir with the thermometer. Read and record the temperature every 15 seconds. Add pure ice as the cubes in the glass melt. Continue stirring and recording the temperature until a constant temperature is reached.

b. Does adding a solute to the water affect the temperature of the ice water? Repeat this experiment adding 4 tablespoons (60 ml) of table salt to ice water in the glass.

c. Rock salt is added to ice in an ice-cream maker to lower the temperature. Is rock salt more effective at lowering the temperature than table salt is? Repeat this experiment using rock salt instead of table salt. Use information about **colligative properties** (properties that depend only on the number of particles in solution) to explain the results.

2. Do solutes affect the freezing of water? Fill two 5-ounce (150-ml) cups with distilled water. Dissolve 1 tablespoon (15 ml) of salt in one of the cups of water. Label the cup containing the salt with the letter S. Set both cups in a freezer. Check the cups every hour for one day; then leave the cups in the freezer overnight.

3. Place the frayed end of a 12-inch (30-cm) cotton string on top of an ice cube. Rub the ice as you press the string against it. The string should cover as much of the surface of the ice as it can and lay flat against the ice. Sprinkle 1 teaspoon (5 ml) of salt over the string. Wait for one minute and gently lift the string. The ice cube should be stuck to the string and it should be possible to suspend the cube by holding the string up. A brief explanation is that rubbing the ice as you press the string onto it melts the ice and that water is absorbed by the string. The salt also melts the ice. The salt then dissolves in the water, producing a salty solution that freezes at a lower temperature than the freezing temperature of water, which is 32°F (0°C). The temperature of the salty water surrounding the string is lower than the freezing temperature of any unsalted water absorbed by the string; thus, the pure water in the string freezes and sticks to the ice cube. Improve upon this explanation and include information about freezing-point depression and the fact that some of the lower-concentration salty water freezes.

Get the Facts

1. A solution has a different boiling point and freezing point than a pure solvent because of the colligative properties of the solution. Colligative properties depend only on the number of particles dissolved in the solvent and not on the nature of the solute or solvent. Find out more about the effects of colligative properties on solutions, including vapor-pressure lowering, boiling-point elevation, and freezing-point depression. How much of a pressure and temperature change can solutes make? Why does salty water have a lower temperature than ice water without the salt? Why are icy sidewalks sprinkled with salt or sand?

2. The freezing and boiling points of a solution can be predetermined with the *molal* freezing-point and boiling-point constants for the solvent. Use a chemistry text to find out about these constants for water. Use the boiling-point constant to calculate the accepted boiling point of solutions containing sugar and salt. Compare the accepted values with your experimental values.

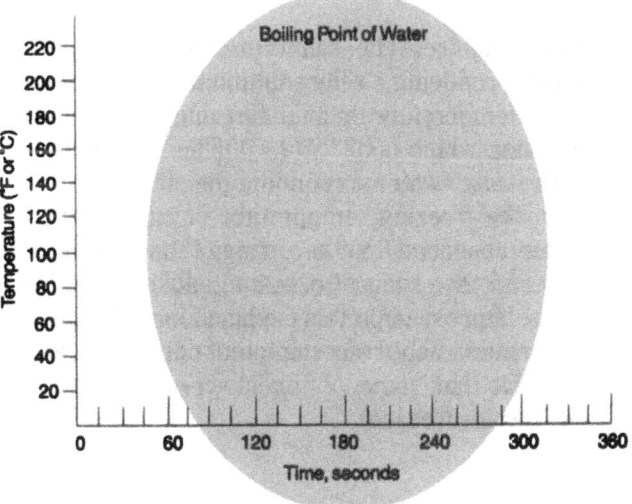

Figure 24.2

Why?

Certain conditions are required to effect a **phase change** of matter from on **physical phase** (gas, liquid, or solid) to another. Heating the water in the pot causes vapor bubbles to form on the bottom and around the sides of the pot where the water is the hottest. As the bubbles rise through the cooler water, they not only cool but also are pressed from all sides by the water molecules, causing them to collapse (see Figure 24.3).

As the water temperature increases, the pressure inside the vapor bubbles increases. Eventually, the **boiling point** is reached (a temperature at which the pressure inside the bubbles equals the atmospheric pressure outside). At the boiling point, the bubbles do not collapse but escape from the surface of the water.

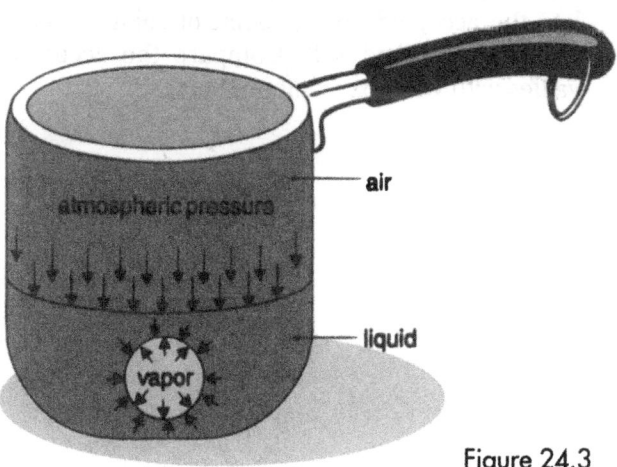

Figure 24.3

During boiling, the temperature of the vapor is the same as that of the liquid. The **kinetic energy** (energy in motion) of liquids is less than that of vapor. Because of the lesser amount of kinetic energy, liquid molecules move less and are more closely bonded together. It takes extra energy to physically break the bonding between liquid molecules and change them into gas. At the boiling point of a liquid, the energy applied by the heating source does not increase the water's temperature but is used in the phase change from liquid to gas.

Try New Approaches

1. Does increasing the heat source raise the temperature of the water? Repeat the experiment using a higher heat setting on the stove.

2a. Does adding a **solute** (substance dissolved in a solution) to the water affect the boiling-point temperature of the water? Repeat the original experiment adding ½ cup (125 ml) of sodium chloride (table salt) to the water.

b. Does using a different solute change the results? Repeat the original experiment adding ½ cup (125 ml) of sucrose (table sugar) to the water. **Science Fair Hint:** Use neatly labeled graphs as part of a display.

Design Your Own Experiment

1a. At what temperature does the **freezing** (the physical change from a liquid to a solid) of water occur? The temperature of a mixture of ice and water will change until the freezing point (or the melting point) is reached. At the freezing point, the temperature of the solution remains constant as energy is used in the phase change. Demonstrate this by filling an ice-cube tray with distrlled water and placing it in a freezer overnight. Fill a drinking glass half full with these pure ice cubes (made with distilled water) and cover the ice with distilled water. Insert a thermometer into the glass of ice water. Set the glass in a large can. Fill the can with ice (this does not have to be pure ice) mixed with 1 cup (250 ml) of table salt.

Gently stir with the thermometer. Read and record the temperature every 15 seconds. Add pure ice as the cubes in the glass melt. Continue stirring and recording the temperature until a constant temperature is reached.

b. Does adding a solute to the water affect the temperature of the ice water? Repeat this experiment adding 4 tablespoons (60 ml) of table salt to ice water in the glass.

c. Rock salt is added to ice in an ice-cream maker to lower the temperature. Is rock salt more effective at lowering the temperature than table salt is? Repeat this experiment using rock salt instead of table salt. Use information about **colligative properties** (properties that depend only on the number of particles in solution) to explain the results.

2. Do solutes affect the freezing of water? Fill two 5-ounce (150-ml) cups with distilled water. Dissolve 1 tablespoon (15 ml) of salt in one of the cups of water. Label the cup containing the salt with the letter S. Set both cups in a freezer. Check the cups every hour for one day; then leave the cups in the freezer overnight.

3. Place the frayed end of a 12-inch (30-cm) cotton string on top of an ice cube. Rub the ice as you press the string against it. The string should cover as much of the surface of the ice as it can and lay flat against the ice. Sprinkle 1 teaspoon (5 ml) of salt over the string. Wait for one minute and gently lift the string. The ice cube should be stuck to the string and it should be possible to suspend the cube by holding the string up. A brief explanation is that rubbing the ice as you press the string onto it melts the ice and that water is absorbed by the string. The salt also melts the ice. The salt then dissolves in the water, producing a salty solution that freezes at a lower temperature than the freezing temperature of water, which is 32°F (0°C). The temperature of the salty water surrounding the string is lower than the freezing temperature of any unsalted water absorbed by the string; thus, the pure water in the string freezes and sticks to the ice cube. Improve upon this explanation and include information about freezing-point depression and the fact that some of the lower-concentration salty water freezes.

Get the Facts

1. A solution has a different boiling point and freezing point than a pure solvent because of the colligative properties of the solution. Colligative properties depend only on the number of particles dissolved in the solvent and not on the nature of the solute or solvent. Find out more about the effects of colligative properties on solutions, including vapor-pressure lowering, boiling-point elevation, and freezing-point depression. How much of a pressure and temperature change can solutes make? Why does salty water have a lower temperature than ice water without the salt? Why are icy sidewalks sprinkled with salt or sand?

2. The freezing and boiling points of a solution can be predetermined with the *molal* freezing-point and boiling-point constants for the solvent. Use a chemistry text to find out about these constants for water. Use the boiling-point constant to calculate the accepted boiling point of solutions containing sugar and salt. Compare the accepted values with your experimental values.

Notes on
Phase Changes

Key Facts:

_____
_____
_____
_____

My Results:

_____
_____
_____
_____
_____
_____
_____

Conclusions:

_____
_____
_____
_____
_____

Results of Try New Approaches:

_____
_____
_____
_____
_____

Notes on Designing My Own Experiment:

_____
_____
_____
_____
_____
_____
_____
_____
_____
_____
_____
_____
_____
_____
_____
_____
_____
_____
_____
_____
_____
_____
_____

Notes on Get the Facts:

_____
_____
_____
_____
_____
_____

25 Crystals
Nature's Jewels

Crystals come in different sizes, shapes, and colors. Some, such as diamonds, are more coveted than others, but they all have their own quality of beauty. By definition, crystals are solids bound by flat surfaces evenly and regularly arranged. Most solids are crystalline, even though their external shapes do not always indicate their orderly internal patterns.

In this project, you will grow crystals that exhibit external structures characteristic of crystals. You will also test the effect of temperature, evaporation rate, and purity of solution on crystal information.

Getting Started

Purpose: To grow sucrose crystals in a gelatin solution.

Materials

½ cup (125 ml) of distilled water

small saucepan

0.25 ounce (7-g) of unflavored gelatin

spoon

stove

1-¼ cups (313 ml) of table sugar (sucrose)

1-pint (500-ml) glass jar

Procedure

1. Pour the water into the saucepan.

2. Sprinkle the gelatin on the surface of the water and let it stand undisturbed for two minutes.

3. Stir the liquid continuously over medium heat until the gelatin in completely dissolved.

4. Add the sugar slowly while stirring.

5. Continue to stir until all of the sugar is dissolved.

6. Remove the saucepan from the heat when the liquid starts to boil.

7. Allow the solution to cool for five minutes.

8. Pour the cooled solution into the jar.

9. Place the jar where it can remain undisturbed for two weeks.

10. Make daily observations of the contents of the jar.

Results

The liquid gels when it reaches room temperature. After two or three days, tiny, clear, glistening **crystals** appear suspended throughout the gel (see Figure 25.1) The crystals grow larger each day and form white, feathery, cloudlike clumps throughout the gel.

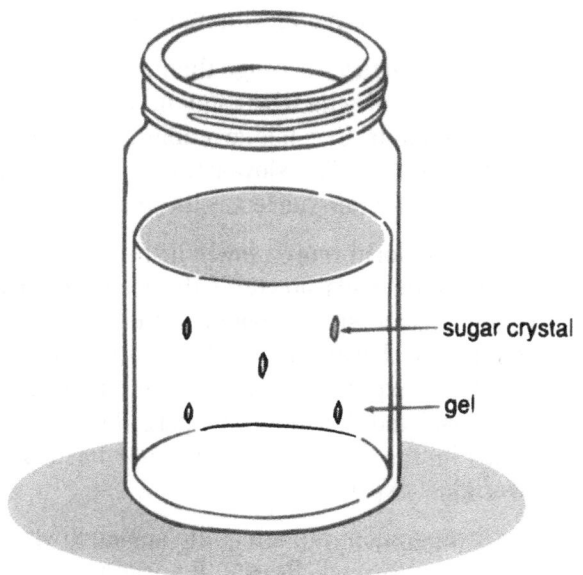

sugar crystal

gel

Figure 25.1

Why?

Solutions contain a solute (what is dissolved) and a **solvent** (what a solute dissolves in). More solute will dissolve in a hot solvent than would dissolve in a cool solvent. In this experiment, extra sugar is dissolved in the water by heating the water. As the solution cools, more solute is dissolved in the solvent than would normally dissolve at the cooler temperature. This cooled solution with excess solute is called a **supersaturated** solution.

As water evaporates from the solution, the solution becomes even more supersaturated. Supersaturated solutions are unstable, and disturbances will cause the microscopic molecules of solute to stick together and form large, visible crystals. The solute is said to "fall out of solution" when the crystals of solute appear in the solution. The first crystals that fall out of solution are too small to be seen, but as more molecules leave the solution, they bind together and form larger and larger crystals. The gel in this experiment allows the crystals to stay suspended; thus, many clusters of crystals can form.

Try New Approaches

1. Will the crystals continue to grow? Observe the size of the crystals in the jar over a two-month period. **Science Fair Hint:** Make weekly observations and draw diagrams of the crystals. Display the diagrams along with photographs.

2. Crystalline particles arrange themselves in positions that require the least amount of energy for their formation. Crystals are able to do this if they form slowly. The slower the crystal forms, the larger and more perfect is its shape.

a. Do crystals form more slowly if the supersaturated solution cools slowly? Is there a difference in the sugar crystal formation if the solution is cooled at a slower rate by insulating the jar? Repeat the experiment placing the jar inside a large, insulated thermos. Compare the crystals formed in this experiment with those formed in the original experiment.

b. Does evaporation rate affect the speed at which crystals are formed? Repeat the original experiment placing the solution in a small-mouthed bottle, such as a soda bottle. The smaller mouth of the bottle slows the evaporation rate of the water from the solution.

3a. Does the gel alter the shape of the sugar crystals? Repeat the original experiment omitting the unflavored gelatin. Tie a paper clip to the end of a cotton string. Cut the string so that it is just long enough to be attached to a pencil and hung inside the jar with the paper clip resting on the bottom of the jar (see Figure 25.2).

b. Repeat this experiment testing the effect that slow cooling has on crystal growth by placing the jar of hot liquid inside a thermos bottle.

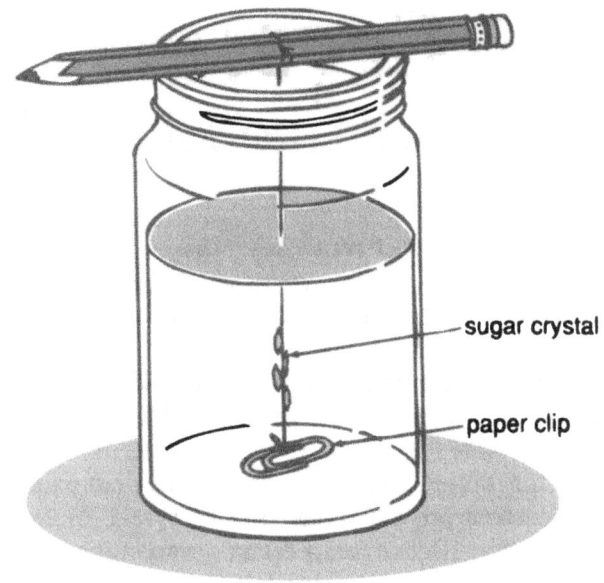

Figure 25.2

c. Determine whether the rate of evaporation affects the growth of crystals in this watery solution. Repeat this experiment and, as before, use a soda bottle to reduce the exposed surface area of the liquid, thus reducing the rate of evaporation.

Design Your Own Experiment

1a. Will sugar crystals form from a solution containing a mixture of different-shaped sugar molecules? Heating a sucrose solution to a high temperature results in the breakdown of some of the sucrose molecules into the smaller sugar molecules of glucose and fructose. An acid, such as cream of tartar, speeds up this breakdown.

Determine whether sugar crystals will grow in a mixture of glucose, fructose, and sucrose by adding together 1-¼ cups (313 ml) of sugar, 1 teaspoon (5 ml) of cream of tartar, and ½ cup (125 ml) of distilled water. Place the mixture in a small saucepan and heat to boiling while stirring continuously. Boil for five minutes. Cool to room temperature before pouring into a clean soda bottle. Hang a cotton string in the bottle, as was done in a previous experiment. Place the bottle where it can remain undisturbed for two or more

Notes on Designing My Own Experiment:

_____
_____
_____
_____
_____
_____
_____
_____
_____
_____
_____
_____
_____
_____
_____
_____
_____
_____
_____
_____
_____
_____
_____
_____
_____

Notes on Get the Facts:

_____
_____
_____
_____
_____
_____

26 The Best Plant Food
Nutrient Differences in Soils

Fertilizer is a natural or synthetic chemical substance or mixture of substances used to add nutrients to soil so as to promote plant growth.

Plants can grow in natural environments without the assistance of commercial fertilizers. The soil seems to provide all the nutrients the plants need. Is there a difference in the nutrients in soils from one location to another?

In this project you will test the effects on plant growth of nutrients leached from soil samples taken from different environments. You will also discover which nutrients are best for plants.

Getting Started

Purpose: To determine whether nutrients leached from potting soil are sufficient for plant growth.

Materials

2-quart (2-liter) food colander

two 2-gallon cooking pots or containers with openings large enough to support the colander

two 2-by-2-foot (60-by-60-cm) pieces of cheesecloth

2 quarts (2 liters) of potting soil

1-quart (1-liter) jar

4 gallons (16 liters) distilled water in plastic jugs

funnel

marking pen

masking tape

8 house plants (same kind and size and each with five or more leaves) in similar containers

Procedure

1. Place the food colander over the opening in one pot so that the rim of the pot supports the colander.

2. Line the inside of the colander with one piece of cheesecloth.

3. Fill the cloth-lined colander with potting soil.

4. Use the jar to pour 2 gallons (8 liters) of distilled water evenly over the soil (see Figure 26.1). *Note:* It will take time for all of the water to drain

through the soil, so add the water slowly. The water that has drained through the soil is called the filtrate.

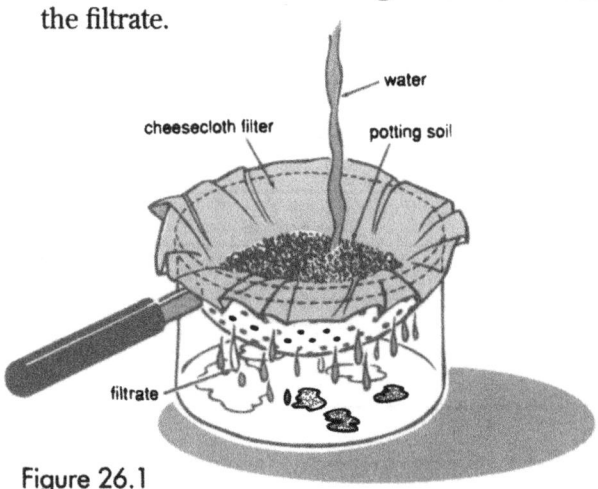
water
cheesecloth filter
potting soil
filtrate

Figure 26.1

5. After all the filtrate has drained through the soil, lift the soil-filled colander and place it over the empty pot.

6. Pour the filtrate through the soil again.

7. Repeat steps 5 and 6 three times. Keep the washed soil.

8. Use the funnel to pour the filtrate into the two empty water jugs. Cap the jugs.

9. With the marking pen and tape, label these jugs "Nutrient Filtrate."

10. Label the remaining jugs of water "Water Only."

11. Fill one of the cooking pots about three-fourths full with tap water.

12. Remove the plants from their containers. Gently shake the soil from their roots and stand the plants in the pot of water to soak off more soil.

13. Remove all the soil from the plant containers, rinse them with tap water, and repot the plants using the washed soil from step 7.

14. Use the marking pen to label four of the plant containers "Water Only" and the remaining four "Nutrient Filtrate."

15. Place all the plants together in a spot where their

environment (temperature, amount of sunlight, and so on) is the same.

16. Keep the soil moist, but not dripping wet, consistently using the liquid indicated on the plant container's label. Add the same amount of liquid to each plant, and water them all at the same time.

17. Observe and record the growth of the plants for four or more weeks.

Results

Your results may vary, but the author found that at the end of four weeks the nutrient-fed plants had stronger stems and greener leaves.

Why?

Many nutrients needed for plant growth are **soluble** (able to break up and thoroughly mix with another substance) in water. Thus, when mixed with water they form an **aqueous solution,** which is a mixture in which one or more substances are dissolved in water. (**Dissolve** means to break into very small particles and mix thoroughly with a liquid.) Nutrients in soil can be removed by a process called leaching. **Leaching** is the removal of soluble chemicals from a material, such as soil, by filtering water through the material so that water-soluble substances are extracted. In this experiment, the cheesecloth acts as a **filter** (material that allows a liquid, but not solids, to pass through) and the liquid passing through the cheesecloth is called the **filtrate** (liquid that passes through a filter). The filtrate collected by leaching is rich in nutrients. Plants grown with this nutrient liquid grow better than plants grown in the washed soil without nutrients. Plants without soil nutrients continue to make food in their leaves by photosynthesis, but photosynthesis alone is not enough to sustain the plants. The nutrients taken in by the roots are necessary for proper growth and maintenance of plant cells. Lack of nutrients results in many problems, including yellow leaves, wilting, thin foliage, small leaves, and generally poor growth.

Try New Approaches

Do all soils contain the same nutrients? Repeat the experiment using samples of soils taken from different locations. Remove any ground covering, grass, and/or plants growing in the soil. Be sure to label the soil samples and make notes of the types of plants growing in the soil and in the general area from which

each sample is taken (see Figure 26.2). This information can be used later to determine the nutrients needed by these plants. Use leached water from the different soil samples to grow plants. Determine which nutrient filtrate is the best for the plants used.

Figure 26.2

Design Your Own Experiment

The types of nutrients in soil can be determined with an inexpensive soil-testing kit. (Your teacher can order this from a plant nursery or a science supply company. See Appendix 10 for a list.) Design an experiment to compare the effect of specific nutrients, such as nitrates and phosphates, on plants. One way is to repeat the original experiment, but test for and identify the presence of some of the nutrients. Record your testing results in a Nutrient Soil Data table like Table 26.1. You may wish to include information in the table about where the soil was found, as shown in Table 26.1.

TABLE 26.1 NUTRIENT SOIL DATA			
Soil Sample	Nitrate Test	Phosphate Test	Soil Location
1	yes	yes	potting soil
2	yes	no	open field with only crabgrass growing in it
3			
4			

Get the Facts

Plants look unhealthy when they do not have proper nutrients. Use a gardening book to determine the nutrients needed for proper plant growth for the plants in your experiments and the symptoms that indicate a deficiency of each nutrient.

NAME

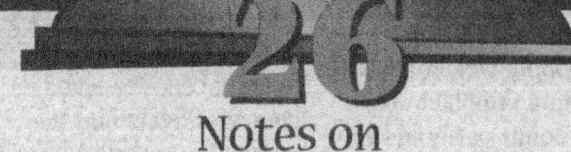

Notes on
The Best Plant Food

Key Facts:

_____
_____
_____
_____

My Results:

_____
_____
_____
_____
_____
_____
_____

Conclusions:

_____
_____
_____
_____

Results of Try New Approaches:

_____
_____
_____
_____
_____

Notes on Designing My Own Experiment:

_____

_____

_____

_____

_____

_____

_____

_____

_____

_____

_____

_____

_____

_____

_____

_____

_____

_____

_____

_____

_____

_____

_____

_____

_____

Notes on Get the Facts:

_____

_____

_____

_____

_____

_____

27 Acid-Base Indicators
Color Changers

Acid/base indicators are chemicals that help you compare or estimate the strength of an acid or base by changing color depending on the pH value of a tested liquid. The lower the pH value, the more acidic the liquid. The higher the pH value, the more basic, or alkaline, the liquid.

In this project, you will make indicators from foods, such as red cabbage and various plants. You will determine the affect of acids and bases on the color of indicators and create color scales for the different indicators. You will also investigate the effect of soil pH on the color of flowers containing an acid-base indicator.

Figure 27.1

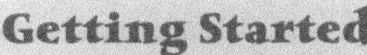
Getting Started

Purpose: To determine the effect of acids and bases on red cabbage indicator.

Materials

pen
masking tape
1-tablespoon (15-ml) measuring spoon
white vinegar, 5%
distilled water
½ teaspoon baking soda
three 10-ounce (300-ml) clear plastic cups
sheet of white copy paper
red cabbage indicator (see Appendix 6)
4 stirring spoons

Procedure

1. Use the pen and tape to label the cups "Acid," "Base," "Neutral."

2. Add 1 tablespoon (15 ml) of vinegar to the Acid cup, 1 tablespoon (15 ml) of distilled water and ½ teaspoon (2.5 ml) baking soda to the Base cup, and 1 tablespoon (15 ml) of distilled water to the Neutral cup. *Note:* Wash the spoon in distilled water and dry it after each use so that you do not contaminate or dilute the contents of the different cups.

3. Set the cups on one of the sheets of white paper so that it will be easier to detect color changes.

4. Add 1 tablespoon of red cabbage indicator to each cup. Using different spoons, stir the contents of each cup. Then observe and record their color in a Red Cabbage Indicator Data table like Table 27.1.

TABLE 27.1 RED CABBAGE INDICATOR DATA	
Solution	Color of Indicator
Acid (vinegar, 5%)	red
Water (distilled)	purple
Base (baking soda)	blue

Results

The red cabbage indicator turns red in vinegar, remains purple in water, and turns blue in baking soda.

Why?

Chemicals are substances made of a combination of **elements** (basic chemical substance of which all things are made). Each element is made of **atoms** (building blocks of matter), and only one kind of atom makes up each specific element. Indicators are natural or synthetic chemical substances that change color in response to other chemicals. **Acid/base** indicators change colors in the presence of acids or bases. **Acids** are aqueous solutions containing hydronium ions—H_3O^{+1} and **bases** are aqueous solutions containing hydroxide ions—OH^{-1}. Generally the characteristics of acids and bases are opposite. The special scale for measuring the acidic or basic nature of a substance is called the **pH scale.** The values on the **pH** (a number used to indicate the acidic or basic nature of a solution) scale range from 0 to 14, with the pH value of 7 being **neutral** (having no acidic or basic properties). Acids have a pH of less than 7 and bases have a pH greater than 7.

Red cabbage contains a chemical that is one of a class of compounds called **anthocyanins** (red pigment found in some foods). The anthocyanin pigment in red cabbage is an acid/base indicator. Vinegar is known to be an acid with a pH of about 2.8 and baking soda is a base with a pH of about 8.5. So from the results of this experiment, cabbage indicator is red at pH of 2.8, purple at pH 7, and blue at pH 8.5. For more information about color changes of anthocyanins in red cabbage due to changes in pH, see pages 10–11 in Dianne N. Epp's *The Chemistry of Food Dyes* (Terrific Science Press: Middletown, Ohio, 1995).

Try New Approaches

Anthocyanins are found in red poppies, blue cornflowers, grapes, beets, blueberries, and apple skin. Repeat the experiment using a different plant and/or fruit to make anthocyanin indicator. How do the colors differ from the colors in the original experiment?

Design Your Own Experiment

1. In the original experiment using red cabbage indicator, only three different testing materials were used. Thus, a pH color scale for the indicator would look like Table 27.2.

TABLE 27.2 PH COLOR SCALE FOR RED CABBAGE INDICATOR

Testing Material	pH	Color of Indicator
Vinegar	2.8	red
Water	7	purple
Baking soda	8.5	blue

Design a more comprehensive pH color scale for red cabbage indicator as well as for one or more of the other food or plant indicators previously made. First test each indicator with testing materials of known pH, such as those in Table 27.3. Make an effort to create a scale with colors for the greatest pH range possible. See a chemistry text for the pH of other materials. *Note:* Mix about ½ teaspoon (2.5 ml) of solid materials with 1 tablespoon of water.

TABLE 27.3 PH DATA

Testing Material	pH
Lemon juice	2.3
Apple juice	3.2
Tomato juice	4.4
Bananas	5.3
Distilled water	7
Baking soda	8.5
Milk of magnesia	10.5
Ammonia	12

2. If you have access to a pH meter, design a pH color scale for red cabbage indicator and/or other food or plant indicators. With assistance from an adult, use an acid stronger than lemon juice or vinegar and a base stronger than ammonia to create a color pH scaled from 0 to 14. As you add drops of acid or base to the indicator, test the liquid's pH, then record the pH and color until you have determined colors for pH 0 to 14.

Get the Facts

The color of some flowers is an indication of the pH of the soil they grow in. Hydrangeas have pink flowers in soil with a high pH (basic) and blue flowers in soil with a low pH (acid). Find out more about soil pH by checking with a plant nursery, school agriculture department, or county agriculture department for information. What can be added to soil to change its pH? **Science Fair Hint:** Determine the range of colors of flowers containing acid-base indicators. Do this by growing plants, such as hydrangeas, in soils of various pH. Your control can be soil with a pH of 7.

Notes on
Acid–Base Indicators

Key Facts:

_____

_____

_____

_____

My Results:

_____

_____

_____

_____

_____

_____

_____

Conclusions:

_____

_____

_____

_____

_____

Results of Try New Approaches:

_____

_____

_____

_____

_____

Notes on Designing My Own Experiment:

_____
_____
_____
_____
_____
_____
_____
_____
_____
_____
_____
_____
_____
_____
_____
_____
_____
_____
_____
_____
_____
_____
_____
_____
_____
_____
_____
_____

Notes on Get the Facts:

_____
_____
_____
_____
_____
_____
_____

28 Enzymes
Chemical Controllers

Advertisements for some detergents proclaim that their product is a miracle worker in a box. What is the ingredient in some detergents that makes them so effective at removing stains?

In this project, you will analyze the chemistry of enzymes, which are the chemicals added to detergents that make them such miraculous cleaners. You will examine the effect of water temperature and pH on the effectiveness of detergent enzymes. You will also determine the effectiveness of enzyme detergents on cleaning different stains.

Getting Started

Purpose: To determine the effect of the enzymes found in a detergent on a protein.

Materials

twelve 1-quart (1-liter) jars
distilled water
marking pen
masking tape
1-tablespoon (15-ml) measuring spoon
powdered laundry detergent without enzymes (Ivory Snow)
stirring spoon
powdered laundry detergent with enzymes (such as Cheer)
12 eggs (fresh, hard-boiled, and peeled)
magnifying lens

Procedure

CAUTION: *When preparing the boiled eggs, wash your hands and any utensils used when handling the uncooked eggs. Raw eggs can contain a harmful bacteria.*

1. Fill the jars three-fourths full with distilled water.

2. Use the pen and tape to label and number the jars "Control-1," "Control-2," "Control-3," and "Control-4"; "With-1," "With-2," "With-3," and "With-4"; and "Without-1," "Without-2," "Without-3," and "Without-4."

3. Add 1 tablespoon (15 ml) of detergent without enzymes to each of the four jars labeled "Without." Stir.

4. Add 1 tablespoon (15 ml) of detergent with enzymes to each of the four jars labeled "With." Stir.

5. Divide the jars into four sets of three jars each. Each set will contain jars labeled "Control," "With," and "Without."

6. Observe and record a preliminary qualitative description of each egg in an Enzyme Data table, like Table 28.1. Then place one egg in each jar.

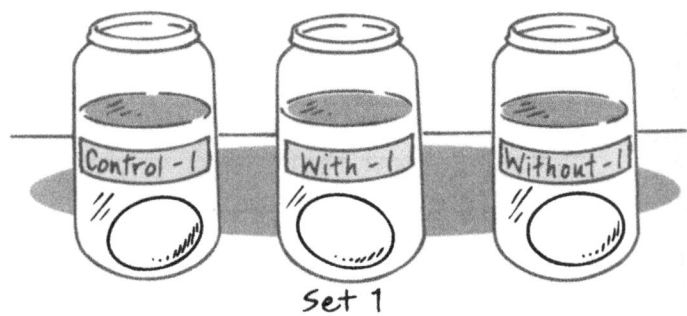

Set 1

Figure 28.1

7. Stand the jars together in an area away from drafts or direct sunlight and with relatively constant temperature.

8. For 7 days, make daily observations of the surface of the eggs. Each day, lift one of the eggs out of its jar, use the magnifying lens for close-up inspection, then replace the egg in its jar. Repeat this procedure for each egg. Make a point not to disturb the surface of the egg any more than necessary during the examination.

Results

The surface of each egg is smooth before being placed in the jars. For each set, as time passes, the surface of the eggs in the detergent with enzymes looks rough and may have a cratered appearance. But those in the detergent without enzymes, as well as the control, remain smooth.

Set	Observation							
1	Start	Day 1	Day 2	Day 3	Day 4	Day 5	Day 6	Day 7
Control-1								
With-1								
Without-1								
2								
Control-2								
With-2								
Without-2								
3								
Control-3								
With-3								
Without-3								
4								
Control-4								
With-4								
Without-4								

TABLE 28.1 ENZYME DATA

Why?

Catalysts are chemicals that change (either speed up or slow down) the rate of a chemical reaction. **Enzymes** are **biological catalysts,** which means they are found in organisms (living things). The substance on which an enzyme acts is called a **substrate.** The specific part of an enzyme that attaches to a substrate is called the **active site.** Enzymes are **proteins,** which are any of a large number of complex **organic compounds** (chemicals containing carbon and found in living organisms) that make up living organisms and are essential to their functioning. Enzymes are different from other proteins because they cause a chemical change, which usually involves the forming or breaking apart of the substrate.

Enzyme action involves three basic steps: First, the enzyme binds to the substrate, then the chemical reaction occurs, and finally the changed substrate is released from the enzyme (see Figure 28.2). Enzymes are usually named after what they bind to (the substrate), followed by the suffix *ase.* So the enzymes in detergent that break apart proteins are called **proteases.** Proteases change larger protein molecules in the egg into smaller molecules, causing the egg to break down. Some proteins, such as hemoglobin, exhibit color because they contain metals. Hemoglobin, for instance, is a blood protein and is red due to the iron attached to it. Proteins are made up of long twisted molecules, which can wind through the fibers of cloth and bind to them. In your wash, the proteases in laundry detergent break down

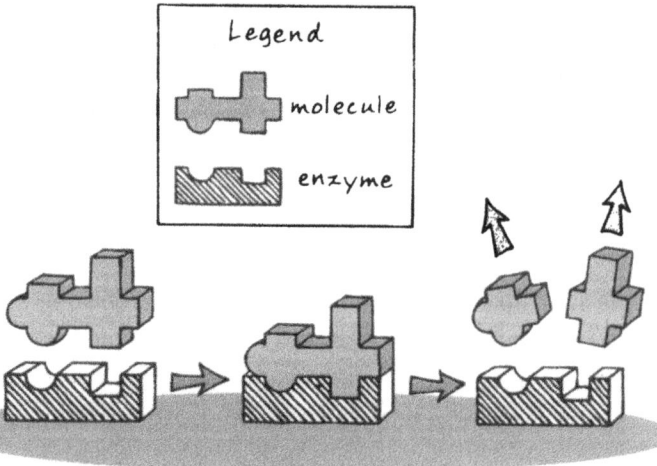

Figure 28.2

protein, such as hemoglobin, thus making it easier for the water and detergent to wash the stains away. Stains that are not caused by colored proteins may not be affected by the proteases.

Try New Approaches

How does pH affect proteases in detergent? Repeat the experiment with the hard-boiled eggs, omitting detergents without enzymes. Instead, have four sets of three jars. For each set 1, label and number the jars "Control-1," "Acid-1," and "Base-1." Fill the jars with distilled water as before. Add 1 tablespoon (15 ml) of detergent with enzymes to jars Acid-1 and Base-1. Use red and blue litmus to test the water in each jar. In the Base-1 jar, if necessary, add 1 teaspoon (5 ml) baking soda at a time so that the pH is high enough to turn the red litmus blue. In the Acid-1 jar, if necessary, add 1 teaspoon (5 ml) of vinegar at a time until the pH is low enough to turn blue litmus red. Repeat preparing sets 2, 3, and 4, then proceed as in the earlier experiment. *Note:* Litmus can be obtained at most teacher supply stores or your teacher can order it from a science supply company. See Appendix 10 for a list of science supply companies.

Design Your Own Experiment

Does temperature affect detergent enzymes? Design an experiment to determine how the temperature of wash water affects stain removal. One way is to use a permanent marker to label twelve 4-inch (10-cm) squares of white cotton cloth. Label 4 pieces "Cold," 4 pieces "Warm" and 4 pieces "Hot." Stain each by placing ⅛ teaspoon (0.72 ml) of yellow mustard in the center of each cloth. Spread the mustard and allow it to dry. Prepare four sets of three 1-pint jars with lids.

Each set will contain jars with cold distilled water, warm distilled water, and hot distilled water. Add 1 tablespoon (15 ml) of detergent with enzymes and 1 stained cloth to each jar of water. With the lids secured on each jar, vigorously shake each jar 5 times every 15 minutes for 1 hour. Then remove the cloth pieces, rinse in cold tap water, and allow to dry. Compare the stains. The cloth pieces in the warm water will be used as a control to compare with the cloth pieces in cold and hot water.

Get the Facts

1. The term *detergent* applies to materials that aid in the removal of dirt or other foreign matter from soiled items. Soap was the main detergent until the 1940s, when synthetic detergents were first developed. Now the term *soap* is usually used to refer to a cleanser made primarily from fat or oil, which doesn't clean well in hard water. Detergent is not made with fat or oil and is chemically active even in hard water. Find out more about the differences and similarities between soaps and detergents. Why are they called surfactants? Explain the hydrophilic and hydrophobic structures of each and how these structures aid them in cleaning. For information, see Carl H. Snyder's *The Extraordinary Chemistry of Ordinary Things* (New York: Wiley, 1997), pp. 323–329.

2. Discover more about the biochemistry of specific enzymes and the proteins they break down. One model that you could use to represent the specific nature of enzymes is the lock-and-key example. Each key fits only one lock, just as each enzyme "fits," or reacts with, only one protein, or a specific class of proteins. Information about the lock-and-key model can be found in a biology textbook.

Notes on
Enzymes

Key Facts:

_____

_____

_____

_____

My Results:

_____

_____

_____

_____

_____

_____

_____

Conclusions:

_____

_____

_____

_____

_____

Results of Try New Approaches:

_____

_____

_____

_____

_____

Notes on Designing My Own Experiment:

_____
_____
_____
_____
_____
_____
_____
_____
_____
_____
_____
_____
_____
_____
_____
_____
_____
_____
_____
_____
_____
_____
_____
_____
_____
_____

Notes on Get the Facts:

_____
_____
_____
_____
_____
_____
_____

29 Triglycerides
Saturated and Unsaturated

Fats and oils are naturally occurring organic compounds called triglycerides. While made of basically the same components, fats and oils can easily be distinguished by the fact that, at room temperatures, oils are liquid while fats are solid. Each molecule of fat or oil is composed of two different types of building-block compounds: glycerol and fatty acids. Both types of compounds are made of atoms of carbon, hydrogen, and oxygen.

In this project, you will study the difference between the molecular structure of saturated and unsaturated triglycerides. You will find out how to determine the iodine number of triglycerides and compare the degree of unsaturation of different oils with different iodine numbers. You will also study the effect of double bonds and length of fatty-acid side chains on the melting point of triglycerides.

Getting Started

Purpose: To use iodine to determine whether a triglyceride is saturated or unsaturated.

Materials

tap water
2-quart (2-liter) cooking pot
heavy-duty aluminum foil
pencil
black permanent marker
5 test tubes
1-tablespoon (15-ml) measuring spoon
safflower oil
eyedropper
tincture of iodine
craft stick or stirring rod
heat mitten
test tube clamp

Procedure

CAUTION: *Keep the iodine out of reach of young children. It is poisonous and is for external use only. Tincture of iodine contains alcohol, which needs to be kept away from an open flame. Iodine will stain skin, clothes, and other materials.*

1. Fill the pot with 2 inches (10 cm) of water.

2. Cover the pot with a piece of aluminum foil that has been folded in half.

3. Use the pencil to make holes in the aluminum foil cover just large enough to stand the test tubes in.

4. Use the marker to label the test tubes "C," "1," "2," "3," "4."

5. Fill each test tube half full with oil. Make sure the height of the oil is the same in each tube.

6. Use the eyedropper to add two drops of iodine to test tubes 1 through 4. Use the craft stick to thoroughly stir the contents of each test tube. Tube C will be the control.

7. Record the initial color of the liquid in each test tube in the row for "0 minutes" in an Oil Data table like Table 29.1

8. Stand the test tubes in the pot by inserting one tube in each hole in the aluminum foil.

9. Turn the stove on to medium heat.

Figure 29.1

10. After 5 minutes, wearing the heat mittens, use the test tube holder to pick up each tube one at a time. Observe the color of the liquid in each tube and record it in the Oil Data table. Return the tubes to the pot of water.

11. Repeat step 10 twice or until no further color change occurs in tubes 1 through 4.

12. Compare the colors of the tubes recorded in the Oil Data table.

Time, min	Color of Oil				
	Test Tube 1	Test Tube 2	Test Tube 3	Test Tube 4	Test Tube C
0					
5					
10					
15					

TABLE 29.1 OIL DATA

Results

The addition of iodine to the pale yellow oil in tubes 1 through 4 turns the oil a pale reddish brown. At first small drops of iodine can be seen suspended throughout the liquid in these tubes. After heating, the reddish color of the liquid in the tubes slowly fades and the liquid returns to its original pale yellow color. The liquid in control test tube C does not change from its original color.

Why?

Molecules are two or more atoms bonded (linked) together. A **bond** is a force holding atoms together. Fats and oils make up a group of naturally occurring organic chemical compounds called triglycerides. **Triglycerides** are organic chemical compounds made of one molecule of **glycerol** (organic alcohol) with three molecules of **fatty acids** (organic acids). Each glycerol molecule has a structure of three carbon atoms with one fatty acid molecule branching from each carbon atom. Fatty acids are so named because they are commonly found in fats and oils. **Organic acids** or organic compounds with one or more acid groups are called the carboxyl group (COOH). Fatty acids have carbon "backbones," usually from 4 to 24 carbons or more, with one carboxyl group at the end. The three acid side chains can be alike or different from one another. Each carbon atom in an acid chain can be attached to another carbon by either a single or double bond. Organic compounds with single carbon-to-carbon bonds are called **saturated,** while **unsaturated** organic compounds have one or more multiple carbon-to-carbon bonds. Saturated fatty acid contains all the hydrogens they are capable of holding, thus the acid is said to be saturated with hydrogen. As the number of double bonds increases, the number of hydrogens attached to the carbons decreases. If the fatty acid has one double bond it is called **monounsaturated.** If there are two or more double bonds, it is called **polyunsaturated.** A **saturated triglyceride** has only saturated fatty acid chains; a **monounsaturated triglyceride** has one or more monounsaturated fatty acid chains; and a **polyunsaturated triglyceride** has one or more polyunsaturated fatty acid chains.

Tincture of iodine contains **elemental** (the natural, noncharged form of an element) iodine (I_2) dissolved in alcohol. When mixed with an unsaturated molecule, such as unsaturated triglycerol, the iodine causes the mixture to look reddish. After heating, the double bonds break and iodine bonds to the carbons. The elemental iodine, along with its color, disappears. (The heat speeds up the process.) The combination of iodine with ethene (an unsaturated organic compound) provides a simple illustration of how iodine bonds with an unsaturated molecule. The product resulting from adding iodine to ethene is colorless 1,2-diiodoethane, as shown in Figure 29.2. However, any remaining amounts of uncombined iodine generally are still present, causing the contents of the test tube to be yellow instead of clear.

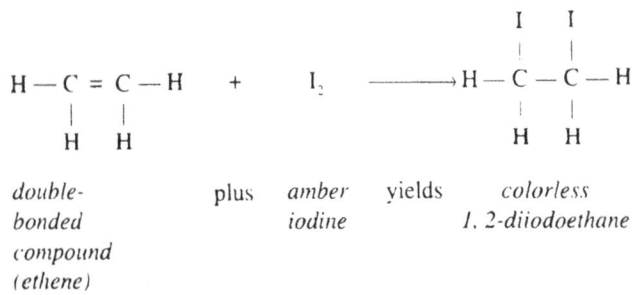

double-bonded compound (ethene)   plus   *amber iodine*   yields   *colorless 1, 2-diiodoethane*

Figure 29.2

Try New Approaches

How does the amount of oil affect additions to carbon-to-carbon double bonds? Repeat the experiment twice, first using more oil. Keep the amount of iodine the same but use 3 tablespoons (45 ml) of oil. Then repeat using less oil, 1 tablespoon (15 ml).

Design Your Own Experiment

The degree of unsaturation of triglycerides is represented by a value known as the **iodine number,** which is the number of grams of iodine that will react with 100 grams of the triglyceride. Table 29.2 gives the iodine number of some common oils. Design an experiment to determine for yourself that oils with a low iodine number are less saturated and will react with a lesser amount of iodine than those with a higher iodine number. Try repeating the original experiment using different oils but start with less iodine, such as one drop. When the oil clears, add another drop. *Note:* Since tincture of iodine is flammable, hold tubes away from the heating source when adding the iodine. Continue to add measured amounts of iodine to each tube of oil until the reddish brown color no longer clears. **Science Fair Hint:** Construct and display a bar graph comparing the number of iodine drops added to each sample of oil.

Get the Facts

1. The number of carbon-to-carbon double bonds in a triglyceride is known as its *degree of unsaturation.* The greater the number of double bonds the greater its degree of unsaturation and the lower the melting point of the triglyceride. How does the length or number of carbons in the fatty acids of a triglyceride affect melting point? For information about triglycerides, see Carl H. Snyder's *The Extraordinary Chemistry of Ordinary Things* (New York: Wiley, 1997), pp. 405–406.

2. *Hydrogenation* is the addition of hydrogen to an unsaturated molecule, which causes a decrease in saturation. If a product is made from a highly unsaturated oil, how does partial hydrogenation of the oil affect its physical and chemical properties? For information, see Snyder's *The Extraordinary Chemistry of Ordinary Things*, pp. 409–410.

TABLE 29.2 IODINE NUMBERS FOR SOME OILS

Oil	Iodine Number
Olive oil	75–95
Peanut oil	85–100
Cottonseed oil	100–117
Corn oil	115–130
Fish oils	120–180
Soybean oil	125–140
Safflower oil	130–145
Sunflower oil	130–145
Linseed oil	170–205

Notes on
Triglycerides

Key Facts:

_____
_____
_____
_____

My Results:

_____
_____
_____
_____
_____
_____
_____
_____

Conclusions:

_____
_____
_____
_____
_____

Results of Try New Approaches:

_____
_____
_____
_____
_____

Notes on Designing My Own Experiment:

_____
_____
_____
_____
_____
_____
_____
_____
_____
_____
_____
_____
_____
_____
_____
_____
_____
_____
_____
_____
_____
_____
_____
_____

Notes on Get the Facts:

_____
_____
_____
_____
_____
_____

30 Vitamin C Content
Analysis of Food by Titration

Vitamin C is required for good health. This vitamin is not produced by your body and must be obtained from foods or vitamin tablets.

In this project, you will use the titration method to determine the amount of vitamin C in various foods and vitamin tablets. You will also determine whether fruit drinks and fruit juices have comparable amounts of vitamin C. Vitamin C's antioxidant properties will be studied, and its part in an oxidation-reduction reaction will be examined. You will also look at the sources and uses of different vitamins.

Getting Started

Purpose: To determine the amount of iodine needed to react with 25 mg of vitamin C.

CAUTION: *Be careful not to allow the vitamin C to touch your skin, in case you have any sensitivity to the chemical.*

Materials

100-mg **ascorbic acid** (vitamin C) tablet
wax paper
hammer
½ cup (125 ml) of distilled water
1-quart (1-liter) jar
spoon
4 baby-food jars

marking pen
masking tape
1-teaspoon (5-ml) measuring spoon
starch solution (see Appendix 6)
sheet of white paper
eyedropper
tincture of iodine

Procedure

CAUTION: *Keep the iodine out of reach of young children. It is poisonous and is for external use only. Tincture of iodine contains alcohol, which needs to be kept away from an open flame. Iodine will stain skin, clothes, and other materials.*

1. Prepare a standard vitamin C solution by:
   - crushing the vitamin C tablet (place it between two sheets of wax paper and hit it gently with a hammer).
   - pouring ½ cup (125 ml) of distilled water into the quart (liter) jar.
   - adding the crushed vitamin C powder to the water in the jar.
   - stirring until the powder dissolves.

2. Pour equal portions of the standard vitamin C solution into four baby-food jars.

3. With the marking pen, write "A," "B," "C," and "D" on pieces of masking tape and tape one label to each jar.

4. Add 1 teaspoon (5 ml) of the starch solution to each jar.

5. Place jar A on the sheet of white paper.

6. Fill the eyedropper with tincture of iodine. Slowly add the iodine in the eyedropper to jar A, counting each drop added (see Figure 30.1). Swirl the jar after each addition of five drops. Continue to add the iodine until the jar's contents remain a blue-black color.

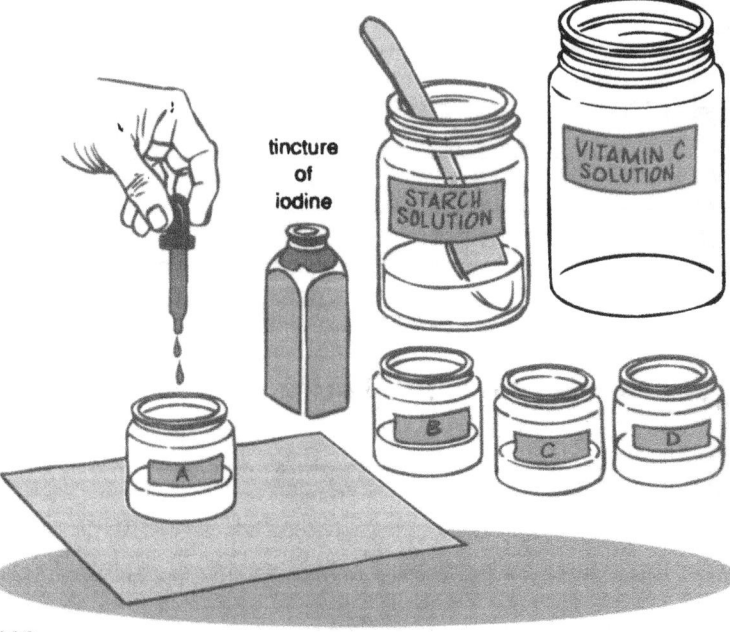

Figure 30.1

7. Record the number of drops used in a Vitamin C Data table, such as Table 30.1.

8. Repeat the procedure using jars B, C, and D.

9. Add the results for the four jars and divide by four to compute the average number of drops of iodine needed to react with the 25 mg of vitamin C in each jar. This number will be used to calculate the concentration of vitamin C in other substances.

TABLE 30.1 VITAMIN C DATA					
	Vitamin C + Starch Solution				
	Jar A	Jar B	Jar C	Jar D	Average
Drops of I_2					

Results

The vitamin C–starch solution is unaffected by the initial drops of iodine, but adding more iodine results in a blue-black solution. The number of drops of iodine needed to react with the 25 mg of vitamin C in the solution will vary with the size of the eyedropper.

Why?

Titration is the process of combining a measured amount of a solution of known concentration with a measured amount of solution of unknown concentration. Tincture of iodine is a mixture of elemental (natural noncharged form of an element) iodine (I_2^0) and ethyl alcohol (C_2H_5OH). A chemical reaction in which electrons are transferred between reactants is known as an **oxidation-reduction reaction,** also called a redox reaction. Originally the term **oxidation** meant the combination of a substance with oxygen and **reduction** meant the loss of oxygen from a substance. Today, scientists extend these definitions to include all transfers of electrons. So oxidation is a loss of electrons or gain of oxygen and reduction is the gain of electrons or loss of oxygen. In the chemical reaction between vitamin C and elemental iodine (I_2^0), vitamin C is easily oxidized (loses electrons), forming **dehydroascorbic acid** and iodine is reduced (gains electrons) forming the **anion** (a negatively charged particle) called **iodide** (I^{-1}).

Elemental iodine reacts with starch to form a complex molecule that has a blue-black color. However, when the iodide anion (the reduced form) is present, the blue-black color is not seen. When starch, vitamin C, and elemental iodine are mixed, the iodine quickly reacts with the vitamin C. When all of the vitamin C has been oxidized by the iodine, additional drops of iodine react with the starch and the blue-black color is seen.

Try New Approaches

Analyze a multivitamin tablet to determine the amount of vitamin C it it. Repeat the experiment substituting a multivitamin tablet for the vitamin C tablet. *Note:* Avoid using a multiple vitamin that contains vitamin E, which may affect the results.

The number of drops of iodine needed to react with the vitamin C mixture depends on the size of the drops, which is determined by the eyedropper. Thus, the same size eyedropper must be used throughout the experiment. Use the following equation and the calculated number of drops of iodine required to react with 25 mg of vitamin C to compute the amount of vitamin C in the test material (the multivitamin tablet). For example, if it takes 50 drops of iodine to react with 25 mg of vitamin C, determine the number of milligrams of vitamin C in a multiple vitamin tablet if 20 drops of iodine reacted with the multiple vitamin.

$$\frac{? \text{ mg of vit. C in multivitamin}}{20 \text{ drops of iodine}} = \frac{25 \text{ mg of vit. C}}{50 \text{ drops of iodine}}$$

$$? \text{ mg of vit. C in multivitamin} = \frac{25 \text{ mg of vit. C} \times 20 \text{ drops of iodine}}{50 \text{ drops of iodine}}$$

$$= 10 \text{ mg of vit. C}$$

Design Your Own Experiment

1. Commercial foods are generally heated when being prepared for packaging. Design an experiment to determine if heating affects the amount of vitamin C in foods. One way is to compare heated with unheated citrus juices. Prepare fresh orange juice by squeezing enough oranges to collect 1 cup of juice. Strain the juice through cheesecloth to separate out any pulp from the orange. Heat half of the juice to boiling, then let it cool to room temperature. Repeat the original experiment twice, first substituting ½ cup (125 ml) of fresh

orange juice for the vitamin C solution, and then using ½ cup (125 ml) of the heated orange juice. *Note:* Both juices need to be at room temperature when you test them.

2. Vitamin C is readily oxidized by combining with oxygen. Design an experiment to determine if storing orange juice in an open container affects its vitamin C content. One way is to store orange juice in two containers, one open (exposed to air, which contains oxygen) and one closed. Keep the containers in a refrigerator so the juice does not spoil. Each week for four or more weeks repeat the original experiment testing the vitamin C content in the juice in both containers. To determine the total amount of juice needed in each container, first decide on the number of days the testing will be done. Then multiply the total number of days by 1 cup (250 ml), the amount of juice needed for each day's testing.

Get the Facts

1. Vitamins contain a variety of complex chemicals. Like ascorbic acid (vitamin C), vitamins have names but are frequently identified by letters, such as A, B, C, D, E, and K. How all vitamins are used by the body is not exactly known, but we do know that many chemical reactions cannot occur in the human body if the proper vitamins are missing. For example, for calcium and phosphorous to be effective in the formation of teeth, vitamin D must be present. Use chemistry and biology texts to find out more about the uses of vitamins. Are chemists, such as Linus Pauling, correct about the role of vitamin C in helping to prevent colds? Is there any evidence to substantiate Pauling's views about the vitamin? What chemical reactions is vitamin C known to be involved in that keep your body functioning normally? What effect does the lack of individual vitamins have on the human body?

2. When skin is exposed to sunlight, the ultraviolet radiation in the light causes *cholesterol* (a fat-related compound) to be chemically converted to vitamin D. Fresh citrus fruits are a good source of vitamin C. Find out more about the natural sources of vitamins. How have the refining and the processing of foods affected their vitamin content?

3. The titration method was used in this project to determine an unknown concentration of vitamin C. Find out more about this process. Use a chemistry text to find the meaning of the following terms: *titration, end point, indicator,* and *standard solution.* You could use this information when preparing an oral presentation and a written report.

4. Combining vitamin C and tincture of iodine results in a redox reaction in which ascorbic acid is oxidized and iodine is reduced. Find out more about the following terms: *redox reaction, oxidized, reduced, oxidizing agent,* and *reducing agent.* What is the difference between the structure of vitamin C (ascorbic acid) molecules and the structure of dehydroascorbic acid, into which the vitamin C is changed as a result of the redox reaction?

Notes on
Vitamin C Content

Key Facts:

_____
_____
_____
_____

My Results:

_____
_____
_____
_____
_____
_____
_____
_____

Conclusions:

_____
_____
_____
_____

Results of Try New Approaches:

_____
_____
_____
_____

Notes on Designing My Own Experiment:

_____
_____
_____
_____
_____
_____
_____
_____
_____
_____
_____
_____
_____
_____
_____
_____
_____
_____
_____
_____
_____
_____

Notes on Get the Facts:

_____
_____
_____
_____
_____
_____

31 Minerals
Chemicals Needed for Life and Good Health

Food minerals are required for life and good health. They consist of inorganic (compounds without carbon and hydrogen) salts and are found in the foods you eat.

In this project, you will examine the functions of different food minerals and their effect on the human body. You will learn about the deposition and growth of calcium crystals within the body and demonstrate reverse calcification of a bone. You will also look at the effect of mineral ion concentration inside and outside body cells.

Getting Started

Purpose: To demonstrate the effect of acids on bones.

Materials

steak knife
uncooked chicken leg
1-quart (1-liter) jar with lid
white vinegar

Procedure

1. Carefully cut as much of the meat away from the chicken leg bone as possible.

2. Examine the flexibility of the bone by trying to bend it with your fingers.

3. Place the cleaned bone in the jar.

CAUTION: Wash your hands after handling the chicken because uncooked chicken can be contaminated with salmonella bacteria.

4. Cover the bone with vinegar.

5. Secure the lid on the jar (see Figure 31.1)

6. After 24 hours, remove the bone from the jar and examine it for flexibility.

7. Replace the bone in the vinegar.

8. Examine the bone for flexibility each day for seven days.

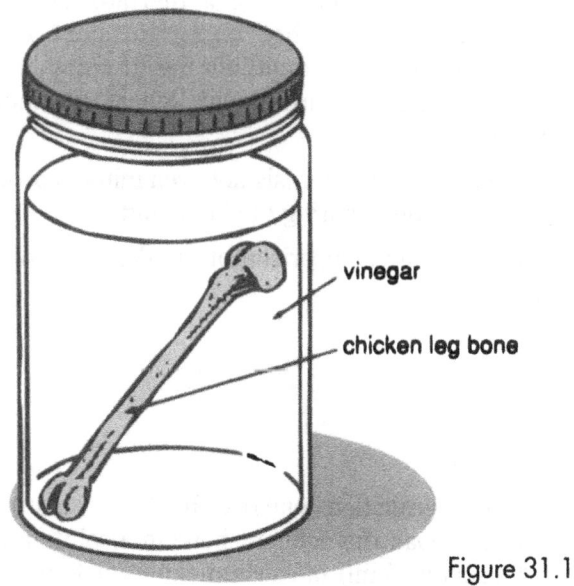

vinegar

chicken leg bone

Figure 31.1

Results

The flexibility of the bone increases daily. At the end of the test period, the bone feels very rubbery.

Why?

During fetal development, strong fibers of protein called **collagen** form a **matrix** (pattern) for bones. The matrix is shaped like the bones but is very flexible. The matrix solidifies by a process called **calcification.** During this process, calcium phosphate or hydroxyapatite, $Ca_{10}(PO_4)6(OH)_2$, is deposited in the fibers of collagen and gives the bones strength and rigidity.

In this experiment, soaking the bone in vinegar leaches out the calcium compounds, resulting in **reverse calcification.** The rubbery bone-shaped form that results is the original collagen "mold" for the bone.

Try New Approaches

Bacteria in your mouth chemically change some of the sugars in food into acid. This acid, like vinegar, reacts with the hydroxyapatite compound in your teeth. Does body temperature speed up the decalcification reaction? Does temperature affect the speed of the reaction between the acid and the calcium compound? Repeat the experiment two times, first storing the jar containing the vinegar and bone in the refrigerator, and then using a thermos bottle of heated vinegar. Heat the vinegar to about 100°F (38°C), pour the hot liquid into the thermos bottle, add the bone, and secure the lid. Reheat the vinegar each day.

CAUTION: Wear heat-resistant oven mitts to protect your hands when pouring the hot liquid.

Make a daily comparison of the flexibility of the bones.

Design Your Own Experiment

1. Can a decalcified bone resorb (absorb again) calcium? Soak the rubbery bone in a solution of 1 teaspoon (5 ml) of calcium oxide (pickling lime from the grocery) and 1 quart (1 liter) of water. Test the flexibility of the bone each day for seven days.

2. Malcolm Bourne, a food scientist at Cornell University, is working on a process of putting the firmness back into cooked vegetables. He lowers the cooking temperature and adds calcium to the food. Bourne claims that the calcium molecules chemically combine with the **pectin** (gluey material that binds vegetables fibers together) in the vegetable. This "salt bridge" holds the vegetable molecules together, resulting in a firmer texture.

   Blanching is the process of cooking vegetables briefly at a high temperature to drive out gases that can cause sour tastes and smells. Blanch 1 quart (1 liter) of fresh string beans by placing them in 2 quarts (2 liters) of boiling water and keeping the water at a boil for five minutes. Remove from the heat, cool, and test their crispness by bending several beans back and forth with your fingers.

   Test Bourne's blanching process by adding ½ teaspoon (2.5 ml) of calcium oxide (pickling lime) to 2 quarts (2 liters) of water. Heat the water to approximately 150°F (66°C). Place 1 quart (1 liter) of fresh string beans in the water for five minutes. Remove, cool, and test for crispness as before. Compare the crispness of these beans with the crispness of the beans in the first blanching process. You could repeat the Bourne process, changing the amount of calcium oxide added to the water.

3. Sodium is the major **cation** (positive chemical particle) in **extracellular fluid** (fluid outside the cell membrane). It is responsible for keeping water in the blood from entering the cells. Potassium is the major cation inside the cell, and it keeps water from leaving the cell. When the concentration of either of these **ions** (charge particles in solution) gets out of balance, water moves into or out of cells until a new balance is achieved. The movement of water through the cell membrane is called **osmosis.**

4. To demonstrate the effect of an increase in sodium ions in extracellular fluids, mix 1 tablespoon (15 ml) of salt in 1 cup (250 ml) of water in a small bowl. Cut three potato slices, each about ¼ inch (6 mm) thick. Check their flexibility by bending the slices back and forth with your fingers. Place the potato slices in the salty water (see Figure 31.2). After 15 minutes, test the flexibility of the potato slices again.

5. To demonstrate the effect of an increase in minerals inside a cell, place the same potato slices in a bowl of fresh water. After 15 minutes, test the slices for flexibility as before.

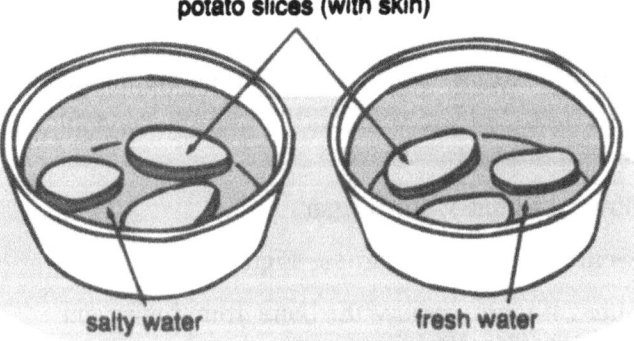

potato slices (with skin)

salty water   fresh water

Figure 31.2

Get the Facts

1. During growth and throughout adult life, bones are remodeled and reshaped. About 20% of adult bone calcium is resorbed by the body each year. Use a nutrition text to find out more about the function of calcium in the human body. How much calcium is deposited each day in adult bones? What is the difference between the calcium compounds in teeth and those in bones? What is calcium's role in blood clotting? What function does calcium play in biological reactions such as the absorption of cobalamin (vitamin B^{12})? What factors affect the absorption of calcium by the body?

2. Phosphorous is a major component of teeth and bones. Find out more about other vital roles that phosphorous plays in body functions. What is its role in the regulation of energy release? How is it involved in the calcification of bones and teeth? What is the food source of phosphorous? How does a deficiency of phosphorous affect the body?

3. Find out more about other necessary body minerals such as magnesium, sodium, potassium, and chlorine. What is the function of each in the body? What is the daily requirement for each? What are the effects of too little or too much of each? What is the source of these minerals?

4. *Plaque* is the thin, adhesive polysaccharide film that covers the enamel layer of teeth. Acids that come from plaque attack the enamel and cause the calcium to be removed, as it was from the chicken bone used in this experiment. Holes form where the calcium is removed. More information about the causes of *dental caries* (tooth decay or cavities) can be provided by your dentist. How do toothpastes prevent the buildup of plaque? What is your composition of a typical *dentifrice* (toothpaste)?

Notes on
Minerals

Key Facts:

_____

_____

_____

_____

My Results:

_____

_____

_____

_____

_____

_____

_____

Conclusions:

_____

_____

_____

_____

_____

Results of Try New Approaches:

_____

_____

_____

_____

Notes on Designing My Own Experiment:

_____
_____
_____
_____
_____
_____
_____
_____
_____
_____
_____
_____
_____
_____
_____
_____
_____
_____
_____
_____
_____
_____
_____
_____
_____

Notes on Get the Facts:

_____
_____
_____
_____
_____
_____

32 Viscosity
A Difference in Stickiness

Liquids move to fill up the space of their containers. This ability to move or flow is a very important physical property of liquids. Viscosity is the measurement of the resistance of fluids to flow.

In this project, you will use the viscosity of common household liquids to make a viscometer, an instrument that measures the flow rate of liquids. The flow rate will be used to calculate the viscosity index, the viscosity of a fluid relative to the viscosity of water, of each testing liquid. You will also determine the effect of temperature on viscosity and look at the binding forces between liquid molecules.

Getting Started

Purpose: To make and use a viscometer to determine the flow rate for a defined volume of water.

Materials

scissors

clear plastic dish detergent bottle with a pull top

marking pen

ruler

modeling clay

glass jar with a mouth slightly smaller than the upper part of the detergent bottle

water

timer

Procedure

1. Cut off the bottom of the detergent bottle.
2. Hold the bottle upside down. With the marking pen, make two straight lines, one about 1 inch (2.5 cm) below the cut-off bottom and the second 4 inches (10 cm) below the first line.
3. Label the first line "Start" and the second line "Stop."
4. Place a ring of clay around the top edge of the jar's mouth.
5. Close the pull top, then stand the bottle upside down inside the jar. Mold the clay ring so that the bottle stands upright, but do not secure the bottle with the clay.
6. Fill the bottle to about ½ inch (1.3 cm) above the start line with cold tap water.
7. Lift the bottle, and with the bottle above the jar, pull the top open.
8. Quickly set the bottle back on the jar (see Figure 32.1).
9. Start the timer when the water level reaches the, start line.
10. Stop the timer when the water level reaches the stop line. Record the time in a Viscosity Data table like Table 32.1.
11. Repeat the procedure three times and average the flow rate of the cold water.

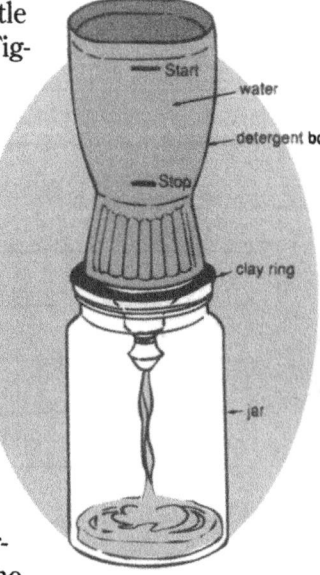

Figure 32.1

	Flow Time, minutes				
TABLE 32.1 VISCOSITY DATA	Test 1	Test 2	Test 3	Test 4	Average
Cold Water					

Results

The author's flow rate for cold water was 39.3 seconds. *Note:* The flow rate will vary depending on the opening and size of the bottle used.

Why?

The amount of time it takes a liquid to flow out of a container depends on its viscosity. The **viscosity** of a liquid is the resistance of the liquid to flowing, because of the friction between the molecules. Viscosity depends on the structure of the liquid molecules. If the molecules are small and simple in structure, as in water, they move past one another quickly. But if they are large and intertwined, as in oil for example, they move slowly past one another. Liquid molecules that slide quickly past one another have a low viscosity; liquid molecules that move more slowly have a high viscosity.

1. Does the temperature of water affect its viscosity? Repeat the experiment twice, first chilling the water in a freezer until its temperature is just above the freezing point, and then using warm tap water.

2. How does the viscosity of other liquids compare with the viscosity of water? The flow or viscosity of a liquid compared with the flow of cold water gives a relative viscosity for the liquid. A number measurement for the relative viscosity of a liquid is called its **viscosity index** (ratio of the rate flow of a fluid to the rate flow of water). Any number less than 1 indicates a lower viscosity than water; a number greater than 1 indicates a higher viscosity than water. Repeat the original experiment using liquids such as oil, dishwashing liquid, honey, and/or syrup. Prepare separate instruments for each liquid tested. Use the flow rate measurements and the following calculation to compute the viscosity index for each liquid. If the flow rate of a liquid is 573 seconds and the flow rate for the same volume of cold tap water is 39.3 seconds, then the viscosity index is determined as follows:

$$\text{viscosity index} = \frac{\text{flow rate of liquid}}{\text{flow rate of water}}$$

$$= \frac{573 \text{ seconds}}{39.3 \text{ seconds}}$$

$$= 14.58$$

This number indicates that the dishwashing liquid is 14.58 times as viscous as is water.

Science Fair Hint: Label and display, in order of viscosity index, photographs of each liquid as it flows from the viscometer.

3a. The viscosity of motor oil is rated by the Society of Automotive Engineers (SAE). The numbers assigned are called "weights" and are not exact viscosity values. Do higher weights of oil indicate a more viscous liquid? Repeat the original experiment using light, medium, and heavy weights of motor oil.

b. How does temperature affect the viscosity of oil? Vary the temperature of the motor oil previously tested and again use the viscometer to determine its flow rate. Place some samples in a freezer overnight to chill the oil. Heat other samples by placing containers of oil in hot tap water. Use a thermometer to determine the exact temperature of each sample tested.

Science Fair Hint: Construct and display a graph showing the relationship between temperature and flow rate.

1a. As the thickness of a liquid increases there is an increase in its viscosity. Design an experiment to compare the viscosity of liquids using their thickness as a means of comparison. One way to compare viscosities is by dropping a glass marble into samples of the liquids. Fill identical slender jars with equal amounts of the liquids. Position the jars in front of a white background to increase your ability to clearly see the results. As you observe the contents of the jars, ask a helper to test two liquids at a time by holding a glass marble in each hand, and holding one hand over each jar. The helper should release both marbles from the same height and at the same time. Observe the movement of the marbles through the liquids. Continue to compare two liquids at a time until you can rate all the liquids in order of their viscosity.

b. Vary the temperature of the liquids. Prepare two jars of each testing sample. Chill one in a refrigerator and heat the second by setting it in a bowl of hot tap water. Measure and record the temperature of each liquid and then repeat the experiment. You could display drawings or photographs of the jars, showing them in order of viscosity.

1. What happens to the *kinetic energy* (energy of motion) of molecules in liquids as the temperature is raised? How does this energy affect the binding forces between the molecules? See a physical science and/or physics text for information about the effect of kinetic energy on viscosity. Use this information to explain the effect of temperature on the viscosity of liquids.

2. Find out more about motor oil. Is there a difference between oil labeled SAE 30 and oil labeled SAE 30W? What does the label 10–W–30 mean? If temperature affects the viscosity of oil, why is it not necessary to change the oil in a car as the seasons change? A good resource is an automotive mechanic.

Notes on
Viscosity

Key Facts:

_____

_____

_____

_____

_____

My Results:

_____

_____

_____

_____

_____

_____

_____

_____

_____

Conclusions:

_____

_____

_____

_____

_____

_____

Results of Try New Approaches:

_____

_____

_____

_____

_____

_____

Notes on Designing My Own Experiment:

_____
_____
_____
_____
_____
_____
_____
_____
_____
_____
_____
_____
_____
_____
_____
_____
_____
_____
_____
_____
_____
_____
_____
_____
_____

Notes on Get the Facts:

_____
_____
_____
_____
_____
_____

33 Carbon Dioxide
Its Production and Uses

Gases, like carbon dioxide, often cannot be seen, felt, or smelled. Yet gases are made of molecules and atoms that chemically react with other substances

In this project, you will test for the presence of carbon dioxide. The physical and chemical properties of the gas will be examined. You will also look at means by which the gas is produced as well as some of its uses.

Getting Started

Purpose: To test for the presence of carbon dioxide when an acid and carbonate compound react.

Materials

glass soda bottle
¼ cup (63 ml) water
⅓ cup of (63ml) vinegar
scissors
ruler
tissue

1 teaspoon (5 ml) of baking soda
baby-food jar with lid
limewater (see Appendix 6)
modeling clay
flexible drinking straw

Procedure

1. Pour ¼ cup (63 ml) of water and ¼ cup (63 ml) vinegar into the bottle.

2. Cut a 3-inch (7.6-cm) strip of tissue

3. Spread the baking soda across the center of the tissue.

4. Roll the paper around the baking soda. Secure the packet by twisting the ends of the paper.

5. Fill the baby-food jar three-fourths full with limewater.

6. Mold a walnut-size piece of clay around the end of the straw, on the end closest to the flexible section. (Do not cover the hole.)

7. Drop the packet of baking soda into the bottle.

8. Quickly plug the bottle's mouth with the clay around the straw. Note: The short end of the straw should be inside the bottle.

9. Hold the jar of limewater near the bottle so that the other end of the straw is beneath the surface of the limewater (See Figure 33.1).

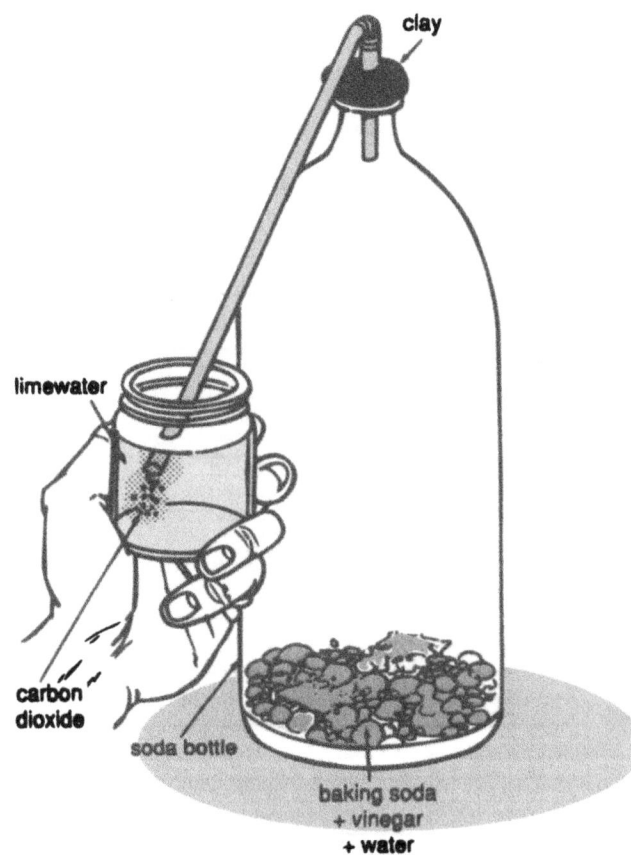

Figure 33.1

10. When the bubbling ceases, observe the limewater.

11. Secure the lid on the jar and allow the jar to stand undisturbed overnight.

12. Observe the contents of the jar.

Results

As bubbles from the straw enter the clear limewater, it turns milky. After standing, the solution looks clear, but there is a thin layer of white solid on the bottom of the jar.

Why?

Baking soda consists of the chemical compound sodium bicarbonate (NaHCO3). Compounds containing carbonate (CO3) react with acids such as vinegar (acetic acid) to produce carbon dioxide gas (CO2). The equation for this reaction is as follows:

$$NaHCO_3 \quad + \quad HC_2H_3O2 \quad \rightarrow$$
sodium   plus   acetic yields
bicarbonate   acid

$$NaC_2H_3O_2 \quad + \quad H_2O \quad + \quad CO_2$$
Sodium   plus water plus carbon
acetate   dioxide

Limewater, Ca $(OH)_2$, is used to test for the presence of carbon dioxide gas because it reacts with carbon dioxide to form the compound calcium carbonate $(CaCO_3)$. The equation for this reaction is as follows:

$$CO_2 \quad + \quad Ca(OH)_2 \quad \rightarrow \quad CaCO_3(s) \quad + \quad H_2O$$
carbon plus calcium yields calcium plus water
dioxide   hydroxide   carbonate

Note: The (s) in the equation indicates that calcium carbonate is insoluble. That is, it does not dissolve. The small white particles of insoluble calcium carbonate temporarily stay suspended in the solution and give it a milky appearance. In time, gravity pulls the calcium carbonate to the bottom of the jar.

Try New Approaches

1. Does the use of a different acid alter the results? Repeat the experiment replacing the vinegar and water mixture with ½ cup (125 ml) of citric acid such as lemon juice or grapefruit juice.

2. Do other carbonated substances produce carbon dioxide when combined with acid? Repeat the original experiment replacing the baking soda with materials such as eggshells or marble chips, which contain calcium carbonate (limestone).

3. Carbon dioxide is a product of the fermentation of sugar. The reaction of the fermentation of sugar is as follows:

zymase
$$C_{12}H_{22}O_{11} \quad + \quad H_2O \quad + \quad yeast \quad \rightarrow$$
sucrose   plus water plus yeast yields

$$4C_2H_5OH \quad + \quad 4CO_2$$
ethyl   plus   carbon
alcohol   dioxide

Zymase is an enzyme (a chemical that changes the rate of a chemical reaction). Demonstrate the production of carbon dioxide by filling the soda bottle half full with water. Add 4 tablespoons (60 ml) of sucrose (table sugar) and ¼ ounce (7 g) of dry yeast. **Science Fair Hint:** Display photographs of the experiment along with the chemical equations of the reactions.

4. Drinking soda contains carbonated water, chemically known as carbonic acid (H_2CO_3). This acid readily decomposes to form water and carbon dioxide $(H_2CO_3 \rightarrow H_2O + CO_2)$. Limewater can be used to test for the presence of carbon dioxide in sodas. Repeat the original experiment replacing the empty soda bottle with a bottle filled with any brand of soda. Speed up the decomposition of the carbonated water by setting the soda bottle in a bowl of warm water. Note: heat the soda bottle only if its cap has been removed.

Design Your Own Experiment

1. Organic fuels, which are compounds containing carbon (wax, for example), produce carbon dioxide when burned. Demonstrate this by burning a candle and collecting and testing the gas produced. Place about 1 inch (2.5 cm) of limewater in the bottom of a 1-quart (1-liter) jar. Wrap the end of a 12-inch (30-cm) wire around the center of a 2-inch (5-cm) long candle. Twist the wire to make a long handle (see Figure 33.2 on the following page). Light the candle and hold the end of the wire to lower the candle into the jar. Cover part of the mouth of the jar with its lid. Allow the candle to burn until the flame goes out. Immediately remove the candle. Secure the lid on the jar and shake the jar vigorously four or five times. A milky solution indicates the presence of carbon dioxide. Display diagrams of a burning candle and of a car, indicating carbon dioxide molecules being emitted from the exhaust of the car and above the flame of the candle.

2. One of carbon dioxide's chemical properties is that it combines with limewater to produce

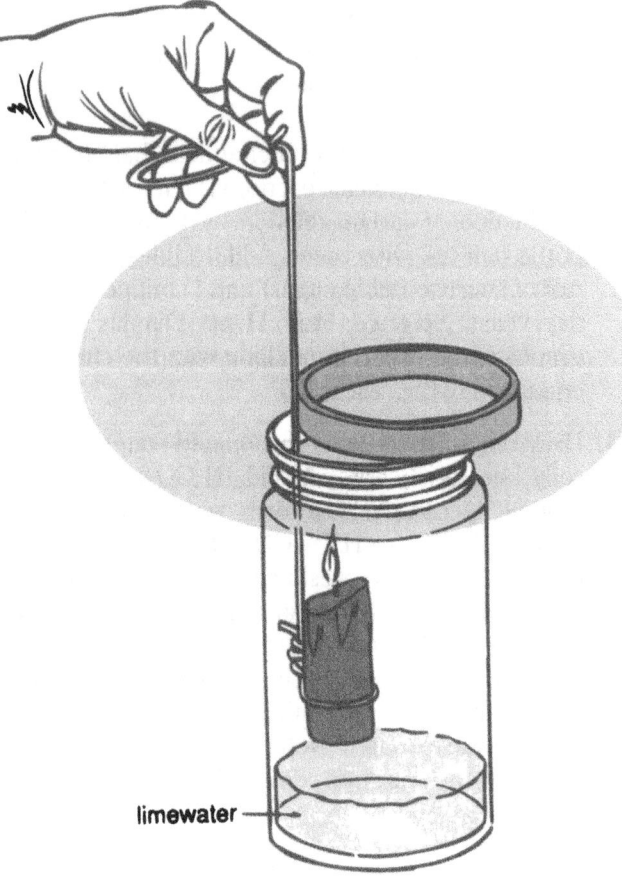

limewater

Figure 33.2

insoluble calcium carbonate. Another important chemical property is that it does not burn or support combustion. Because of this property, carbon dioxide is used in fire extinguishers. Demonstrate this by wrapping a 12-inch (30-cm) wire around a 1-inch (2.5-cm) long candle. Light the candle and hold the end of the wire to lower the candle into a 1-quart (1-liter) jar. In a small-mouthed gallon jug mix together 1 teaspoon (5 ml) of baking soda, ¼ cup (63 ml) water, and ¼ cup (63 ml) of vinegar. When the fizzing stops, hold the mouth of the gallon jug over the mouth of the jar and slowly tilt the jug. Do not allow any liquid from the jug to enter the jar. The invisible, heavier-than-air carbon dioxide gas flows into the jar and extinguishes the flame. Display photographs showing the sequence of events in this reaction along with pictures of carbon dioxide fire extinguishers.

3. Respiration is the chemical process by which animals convert food into energy. Carbon dioxide is a by-product of this reaction and is expelled by the lungs. Test for this gas in your breath by exhaling through a straw into a soda bottle half filled with brom thymol blue solution (see Appendix 6). Brom thymol blue is an indicator that turns yellow in the presence of an acid. Carbon dioxide plus the water in the indicator produces carbonic acid resulting in a green to yellow color depending on the amount of carbon dioxide present.

Get the Facts

1. The temperature of the earth is kept warm due to gases, such as carbon dioxide, in the atmosphere. The atmospheric gases trap warmth from the sun, just as glass traps warmth in a greenhouse. For this reason, this warming of the earth is called the *greenhouse effect*. Scientists think the earth's atmosphere is getting warmer because of the increase of carbon dioxide production. Find out more about the greenhouse effect. The burning of fossil fuels is considered the major cause of the increase of carbon dioxide, but other factors contribute to the increase of this gas. What are they? If the earth's atmosphere is getting warmer, how much warmer is it? What effect does this extra heat have on the earth's environment? Can the warming be stopped?

2. Limestone caverns and the stalagmites and stalactites in these structures are formed from the combination of carbon dioxide and limewater in the soil. Find out more about the chemical reactions involved in the formation of limestone.

3. Find out more about the uses of carbon dioxide such as the following:
   - carbonated drinks
   - leavening agent in baking
   - photosynthesis reaction
   - stimulus for the nerve controlling the diaphragm
   - agent in baking soda and washing soda made by the Solvay process
   - solid carbon dioxide, dry ice, a refrigerant

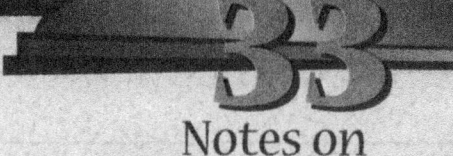

Notes on
Carbon Dioxide

Key Facts:

_____
_____
_____
_____

My Results:

_____
_____
_____
_____
_____
_____
_____

Conclusions:

_____
_____
_____
_____

Results of Try New Approaches:

_____
_____
_____
_____
_____

Notes on Designing My Own Experiment:

_____
_____
_____
_____
_____
_____
_____
_____
_____
_____
_____
_____
_____
_____
_____
_____
_____
_____
_____
_____
_____
_____
_____
_____
_____
_____

Notes on Get the Facts:

_____
_____
_____
_____
_____
_____

Notes on Designing My Own Experiment:

_____
_____
_____
_____
_____
_____
_____
_____
_____
_____
_____
_____
_____
_____
_____
_____
_____
_____
_____
_____
_____
_____
_____
_____

Notes on Get the Facts:

_____
_____
_____
_____
_____
_____
_____

35 Minerals

Distinguishing Physical Characteristics of Minerals

The Earth's lithosphere is the solid part of Earth consisting of the crust and the upper mantle extending to a depth of about 60 miles (96 km). The lithosphere is made up mostly of minerals, which are naturally occurring inorganic solids with a definite chemical composition and physical characteristics.

In this project, you will learn a method to determine the specific gravity of a mineral and discover how heft can be used as a method of mineral identification. The difference between cleavage and fracture will be demonstrated. You will use minerals and everyday objects to demonstrate the hardness of minerals. You will learn how to test minerals for their streak and luster. All of this will then be used to prepare a display of minerals that represents their characteristics.

Getting Started

Purpose: To determine the specific gravity of a mineral.

Materials

2-quart (2-liter) bowl
tap water
24-inch (60-cm) piece of string
fist-size sample of quartz, or any mineral of comparable size
metric spring scale

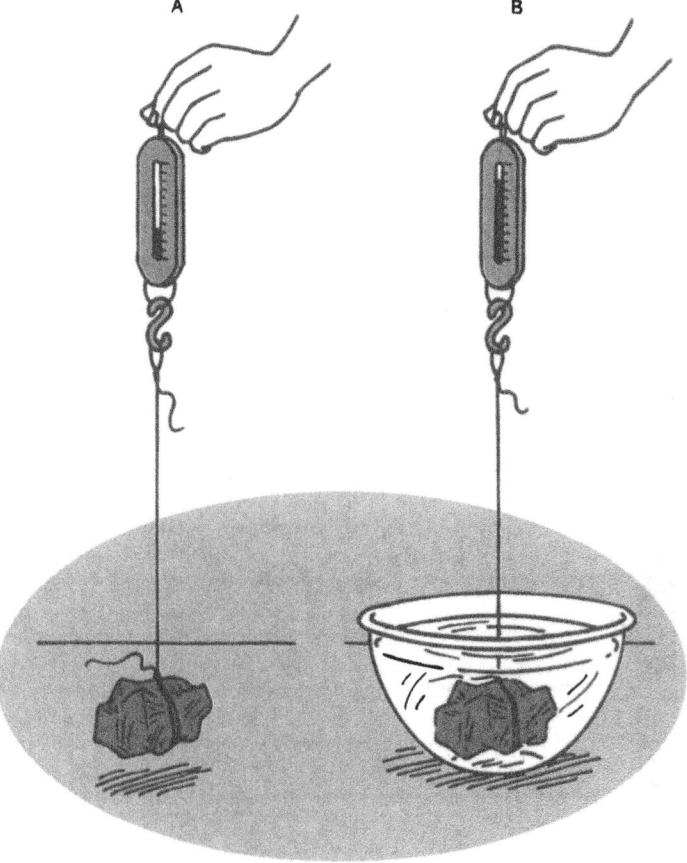

Figure 35.1

Procedure

1. Fill the bowl about three-fourths full with water.

2. Tie the string around the mineral and make a loop in the free end of the string.

3. Place the loop on the scale hook, suspend the mineral in air, and measure the mass of the mineral in grams (see Figure 35.1A). Record this as mass 1 (M_1).

4. With the mineral hanging from the scale, lower the mineral into the water in the bowl (see Figure 35.1B). Do not allow the mineral to rest on the bottom or sides of the bowl. Record this as mass 2 (M_2).

5. Determine the mass of the water **displaced** (pushed aside) by the mineral by calculating the absolute difference between M_1 and M_2. Record the answer as mass 3 (M_3).

Example:

$$M_3 = M_1 - M_2$$
$$= 530 \text{ g} \div 330 \text{ g}$$
$$= 200 \text{ g}$$

6. Use the following example to determine the mineral's **specific gravity (sp. gr.)** (the ratio of the mass of an object in air to the mass of the water displaced by the submerged object):

Example:

$$\text{sp. gr.} = M_1 \div M_3$$
$$= 530 \text{ g} \div 200 \text{ g}$$
$$= 2.65$$

Results

A method for determining the specific gravity of a mineral is used. The specific gravity of the mineral in the example is 2.65.

Why?

Most of the elements in the Earth's crust occur as minerals. A **mineral** is a single solid element or compound found in the Earth and has four basic characteristics: (1) it occurs naturally; (2) it is **inorganic** (not made from living things); (3) it has a definite chemical composition, meaning that it is not a mixture, but is made of the same substance throughout; and (4) it has a **crystalline structure** (atoms are arranged so that they form a particular geometric shape).

One way to distinguish one mineral from another is by comparing the specific gravities of the two minerals. To calculate the specific gravity of a mineral, divide the mass of the mineral by the mass of the water displaced by the mineral. Specific gravity tells how many times heavier the mineral is than water. The mineral in the example is 2.65 times as heavy as the same volume of water. Most minerals have a specific gravity from 2 to 5. Since every mineral has its own particular specific gravity, this characteristic can be used as a clue to the identity of a mineral.

Specific gravity is not always calculated with a scale. Instead, the heft of minerals is often used in identification. **Heft** is a subjective measurement. You measure heft by picking up minerals of equal volume and comparing their weights. Gold and pyrite are minerals that look alike, but the specific gravity of gold is 19.3 and the specific gravity of pyrite is 5.0. Thus, even though heft is not an exact measurement, the heft of gold is easily determined to be greater than that of pyrite.

Try New Approaches

Would the size of a mineral sample affect its specific gravity measurement? Repeat the experiment two times. First, use a larger piece of quartz; then, use a smaller sample of the mineral. **Science Fair Hint:** use the mineral samples and their calculated specific gravities as part of a project display.

Design Your Own Experiment

1. **Cleavage** is a physical characteristic of a mineral which is the tendency to break along a flat surface called a **cleavage plane.** Minerals that break easily and cleanly are described as perfect. Less-clean breaks are described as distinct, indistinct, or none. Minerals can have cleavage in many different directions. Cleavage in one direction is called **basal cleavage.** Muscovite, the most common form of mica, is a mineral that has perfect basal cleavage. Examine a piece of muscovite. You should be able to peel the layers off with your fingers.

    Remove a thin layer of mica and break it in half. Observe the broken edges. This is a **fracture** (an irregular break) which is the way a mineral breaks when it doesn't split along a cleavage plane. Fracture surfaces are described as uneven, **conchoidal** (curved), **hackly** (jagged), or **splintery** (small, thin, and sharp or fibrous surface). Use a rock and mineral handbook to identify the type of fracture that muscovite has.

    Collect and display minerals having different types of cleavage and fractures. Use the handbook to find out more about how minerals break and to discover the identity of the cleavage and fracture type for each mineral specimen.

2a. **Hardness** is a physical characteristic of a mineral, which is the measure of its resistance to being scratched. The scale of hardness from 1 to 10 was devised in 1822 by Frederick Mohs (1773–1839), a German chemist. He arranged 10 common minerals from the softest to the hardest, giving the softest mineral, talc, the number 1, and the hardest mineral, diamond, the number 10. Minerals with higher Mohs' numbers will scratch those with lower numbers. Determine which has a higher Mohs' number, muscovite or quartz. First try to scratch the quartz with the muscovite. Then, try to scratch the muscovite with the quartz. Find the list of minerals for Mohs' scale of hardness in a rock and mineral handbook. Prepare a display showing specimens and/or pictures of the minerals for each Mohs' number.

b. Everyday objects such as the following can be used to represent minerals' hardness:

fingernail	2½
copper coin (penny)	3½
paper clip	4½
pocketknife blade	5½
glass	6
sandpaper	7
steel file	7½

Make and display a chart of the Mohs' scale, similar to the one shown in Figure 35.2.

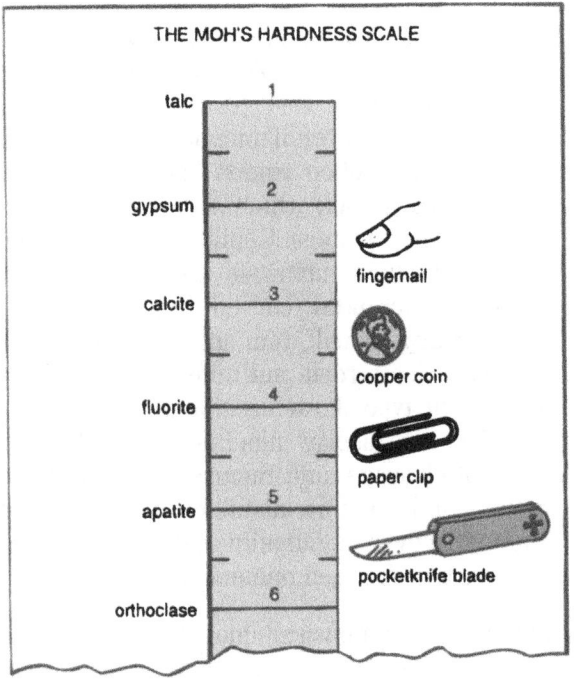

Figure 35.2

3. **Streak** is a physical characteristic of a mineral which is the color of the powder left when the mineral is rubbed against a rough surface that is harder than the mineral. An unglazed porcelain tile, called a **streak plate,** has a hardness of about 7 and is used to determine the streak of minerals with a hardness less than 7. Determine the streak of hematite by trying to scratch the plate with the edge of the hematite. Rub your finger over the powder left on the plate by the hematite and determine its streak color. Note that a mineral's color is not always its streak color. *Note:* A **streak test** is a test in which the streak of a mineral is determined.

4. All minerals have the physical characteristics of specific gravity, crystalline structure, cleavage, fracture, hardness, and streak. Use the previous tests to determine these characteristics of minerals you wish to display. For hints on how to label and display your minerals, see *Janice Van-Cleave's Rocks and Minerals* (New York: Wiley, 1996), pp. 80–83.

Get the Facts

Other than the previously listed physical characteristics, minerals have characteristics such as *luster* (the way a mineral reflects light) and color, and some have magnetism and fluorescence. A rock and mineral handbook is a good source to find out more about these characteristics. What determines a mineral's color or luster? What elements must be present in a mineral for it to have magnetic properties? What is phosphorescence?

Notes on
Minerals

Key Facts:

_____

_____

_____

_____

My Results:

_____

_____

_____

_____

_____

_____

_____

_____

Conclusions:

_____

_____

_____

_____

Results of Try New Approaches:

_____

_____

_____

_____

_____

Notes on Designing My Own Experiment:

_____
_____
_____
_____
_____
_____
_____
_____
_____
_____
_____
_____
_____
_____
_____
_____
_____
_____
_____
_____
_____
_____
_____

Notes on Get the Facts:

_____
_____
_____
_____
_____
_____

36 The Rock Cycle

Processes That Change One Rock Type into Another

Volcanic rocks and fire rocks are common names for igneous rocks. These solidified masses are, as their names imply, the results of great temperatures within the Earth. Igneous rock is one of a trio of rock types—sedimentary, metamorphic, and igneous. Through different processes, each rock type can be changed into one of the other types in the trio. This process of change is called the rock cycle.

In this project, you will study and model the texture of different igneous rocks. The metamorphism of porphyritic rock (a kind of igneous rock) into foliated metamorphic rock will be demonstrated. You will also examine the relationship between the three rock types and model their transformation from one type to the other.

Getting Started

Purpose: To model the difference between a porphyritic rock and other types of igneous rocks.

Materials

two walnut-size pieces of blue modeling clay
two walnut-size pieces of red modeling clay

Procedure

1. Break one red clay piece into four relatively equal size pieces.

2. Roll the four small pieces into balls.

3. Repeat steps 1 and 2 with one blue clay piece.

4. Lay the eight small balls in two rows next to each other, alternating the colors of the balls in the rows.

5. Gently press the clay balls just enough so that they stick together but retain as much of their shape as possible.

6. Break the other large red clay piece in half. From one half, form two relatively equal size balls, and from the other half form four relatively equal size balls.

7. Repeat step 6 with the remaining large blue clay piece.

8. Lay the twelve small balls in two rows next to each other, alternating the colors (and sizes) in the rows.

9. Repeat step 5.

10. Compare the appearance of the two clay rolls (see Figure 36.1).

Results

One of the clay rolls has large balls of clay pressed together. The second has large and small balls.

Figure 36.1

Why?

Rock is a solid, coherent **aggregate** (single mass) of one or more minerals. Rocks produced by the cooling and solidifying of molten rock are called **igneous rocks. Magma** (molten rock under the Earth's surface) at great depths cools slowly, and during this cooling process, large mineral crystals form. Igneous rocks that form within the crust and contain large uniform interlocking crystals are called **intrusive igneous rocks.** The **texture** of rocks is determined by the size of the mineral **grain** (hard particles) making up the rock. Intrusive igneous rocks are **coarse-grained** (having large hard particles). In this experiment, the clay roll made with large clay balls represents a coarse-grained intrusive igneous rock.

In **porphyritic rock,** like other types of intrusive igneous rock, large crystals form from magma cooling at great depths beneath the Earth's surface. However, during the formation of this rock, the magma is pushed to the surface before it completely hardens. There the final cooling occurs rapidly, producing small crystals. Thus, porphyritic rock contains two or more different sizes of interlocking crystals and can be said to have varied grain sizes. The clay roll with the large and small clay balls represents a porphyritic rock.

201

Try New Approaches

Lava (molten rock from within the Earth that reaches the Earth's surface) cools quickly, producing rocks with small crystals or no crystals. Igneous rocks formed by the cooling of lava are called **extrusive igneous rocks.** Extrusive rocks are **fine-grained** (having very small hard particles) or have a **glassy** (smooth glasslike) texture. Repeat the experiment using only small balls of clay.

Design Your Own Experiment

1. **Metamorphism** is the change in structure, appearance, and composition of a rock in the solid state within the Earth's crust as a result of changes in temperature and/or pressure. **Regional metamorphism** occurs when large areas of rock are changed by pressure and heat, such as in mountain formation. What happens to grain arrangement during regional metamorphism when pressure is applied from one direction? Place a piece of paper on a table and position one of the clay rolls from the original experiment on the paper. Cover the clay with a second sheet of paper. Place a rolling pin on the top paper above one end of the clay roll. Pressing down firmly, roll the rolling pin across the clay roll. Repeat using the second clay roll.

   In nature, great pressure on rocks causes the temperature to rise. Together, the heat and pressure changes cause metamorphism, which produces **metamorphic rock.** Pressing the clay models of igneous rock represents the formation of **foliated metamorphic rock** (striped-looking metamorphic rock with grains arranged in parallel bands). **Science Fair Hint:** Make a display to show the clay rolls before pressure was applied to represent igneous rock and after pressure was applied in one direction to represent the resulting foliated metamorphic rock.

2a. Rocks come from other rocks. Igneous rock forms when sedimentary or metamorphic rock melts, cools, and solidifies. **Sedimentary rock** is made from **sediments** (materials deposited by water, wind, or glaciers) of metamorphic or igneous rocks that are **compacted** (packed together) and **cemented** (stuck together). Metamorphic rock forms when igneous or sedimen-

tary rock is changed by metamorphism. This never-ending process by which rocks change from one type to another by a series of processes involving heat, pressure, melting, cooling, and sedimentation is called the **rock cycle.** Draw and display a diagram similar to the one shown in Figure 36.2 to represent the rock cycle. Note that sedimentary rocks are placed at the top of the diagram because these rocks are formed when the other rocks are lifted to or near the top of the Earth's surface, while the formation of metamorphic and igneous rocks is generally below the Earth's surface.

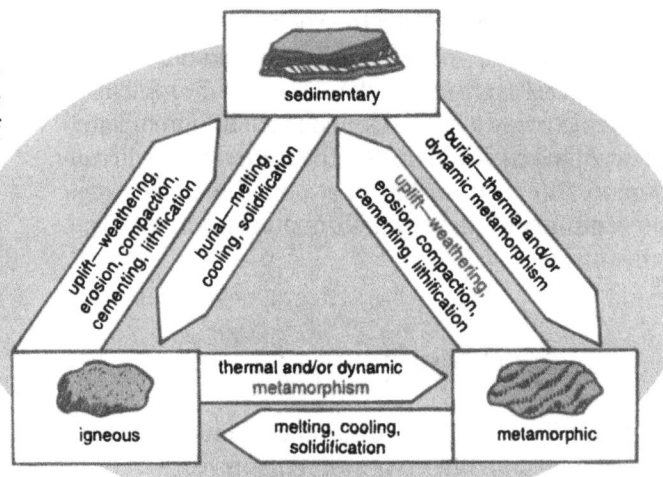

Figure 36.2

b. Use rock samples to prepare a display representing the rock cycle. Use found rocks or rocks purchased at rock and mineral shops or ordered from science catalogs. See Appendix 8 for a list of catalog suppliers and stores selling rocks and minerals.

Get the Facts

Thermal metamorphism includes changes due to heat. *Recrystallization* (enlargement of minerals) is one example of thermal metamorphism. Another example is *contact metamorphism,* which occurs when hot magma intrudes (penetrates) into the rock with which it comes in contact. How large an area is affected by contact metamorphism? What is a metasomatic change? How do contact and regional metamorphism compare? What are the names of different rocks before and after these metamorphic changes? To find out more about metamorphism, see John Farndon's *How the Earth Works* (New York: Reader's Digest Association, 1992), p.83.

Notes on
The Rock Cycle

Key Facts:

_____
_____
_____
_____

My Results:

_____
_____
_____
_____
_____
_____
_____
_____

Conclusions:

_____
_____
_____
_____

Results of Try New Approaches:

_____
_____
_____
_____
_____

Notes on Designing My Own Experiment:

_____
_____
_____
_____
_____
_____
_____
_____
_____
_____
_____
_____
_____
_____
_____
_____
_____
_____
_____
_____
_____
_____
_____
_____

Notes on Get the Facts:

_____
_____
_____
_____
_____
_____

Soil Texture
Effects of Regolith Size

A great part of the Earth's crust is covered with plant growth. The survival of theses plants depends on the physical and nutritional support they receive from a mixture of particles of weathered rock and humus called soil.

In this project, you will learn the differences between coarse-, medium-, and fine-textured soils. How the texture of soil and the shape of its particles affects a soil's porosity will be determined. You will also examine the relationship between soil texture and permeability. Soil profile and types will be studied.

Getting Started

Purpose: To determine the texture of soil.

Materials

garden trowel

1-quart (1-liter) bowl

marker

masking tape

three identical 1-pint (500-ml) plastic transparent jars

newspaper

colander with large holes

large fine-mesh strainer

Procedure

1. Select a spot with soil, such as near a tree or where plants are growing. A bare soil area in a garden is also acceptable. Ask for permission to remove about 1 quart (1 liter) of soil.

2. Use the trowel to fill the bowl with soil.

3. Use the marker and tape to number the jars "1," "2," and "3."

4. Lay the newspaper on a table.

5. Spread the soil on the newspaper and pick out any live animals and parts of dead animals and plants, and return them to where the soil was collected.

6. Pour the soil into the colander, and shake the colander over the newspaper until no more particles fall through the holes in the colander (see Figure 37.1).

7. Put the particles left in the colander into jar 1.

8. Pour the particles on the newspaper into the fine-mesh strainer. Shake the strainer over the newspaper until no particles fall through.

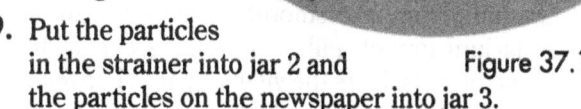

9. Put the particles in the strainer into jar 2 and the particles on the newspaper into jar 3.

Figure 37.1

10. Compare the amount of material in each jar.

Results

The soil is separated into three sizes of particles—large, medium, and small. The amount of material in each jar will vary with different soil samples.

Why?

Soil is the top layer of the regolith (loose particles including soil, that cover Earth's surface) that supports plant growth. Soil composition includes small rock particles, **humus** (decayed animal and plant matter), air, and water. Soil provides support as well as nutrients for plants.

Not all soils are alike. Soil composition can vary by the type of rock and the amount and composition of humus it contains. Most soil contains particles of varying size. In this experiment, you separate the particles. Coarse-grained particles, as in jar 1, are larger than **medium-grained** particles, as in jar 2. Fine-grained particles, as in jar 3, are smaller than the other two particle types. The texture of soil depends on which type of particles predominates in the soil. For example, if there are more particles in jar 3, then your soil sample would be considered fine-textured.

Try New Approaches

1a. Soil texture can be estimated by rubbing it between your fingers. Determine the feel of different textures, then collect a variety of soil samples from very coarse to very fine.

b. Compare the amount of the different-size particles in each sample collected in the previous experiment by repeating the original experiment. **Science Fair Hint:** Use photographs or diagrams of the jars of particles to represent the comparison of particles in different soil textures.

Design Your Own Experiment

1. **Porosity** is the percentage of a material's volume that is **pore space** (small, narrow spaces between particles in materials). How does the shape of soil particles affect porosity? Compare rounded and angular-shaped objects of comparable size. For example, compare smooth rocks or beads with aquarium gravel. Fill a measuring cup with the round objects. Note the visible spaces between the objects. Measure the volume of the spaces by filling the cup to the 1-cup (250-ml) mark with a measured amount of water. Repeat, replacing the rounded objects with the angular-shaped objects. Repeat again, using a mixture of equal parts of the rounded and angular objects.

2. **Permeability** is the measure of how easily a solid allows a **fluid** (any gaseous or liquid material that can flow) to pass through it. If a fluid, such as water, flows quickly through soil, the soil is said to have high permeability. How does texture relate to the permeability of soil? Use the point of a pencil to make six equal-size holes in the bottom of three 9-ounce (270-ml) paper cups. Cut circles from coffee filters to fit in the bottom of each cup. Label the cups "course," "medium," and "fine." Mark a line 2 inches (5 cm) from the bottom of each cup. Fill to this line with samples of the three soil textures corresponding to the labels on the cups. Lay two pencils parallel to each other across a measuring cup. Secure the pencils to the measuring cup with tape, then set the paper cup of coarse soil on the pencils, making sure that the pencils do not cover the holes in the cup (see Figure 37.2). Ask a helper to time you as you pour 100 ml of tap water into the cup. Stop timing when no more water drains from the cup. Record the draining time in a Soil Permeability Data Table like Table 37.1. Measure the amount of water that drained from the paper cup into the measuring cup. Record the amount of water in milliliters. Calculate the drainage rate by dividing the amount of water drained by the time it took to drain. Record this answer. Repeat the procedure

with the other soil samples. The most permeable soil is the one with the highest drainage rate.

Soil Type	Draining Time	Amount of Water Drained	Drainage Rate
course-textured			
medium-textured			
fine-textured			

TABLE 37.1 SOIL PERMEABILITY DATA

Figure 37.2

Get the Facts

1. A cutaway section of Earth would reveal a soil profile made up of layers called *horizons*. Mature soils have three basic horizons. What is the composition of each horizon? For information about horizons, see David Lambert's and the Diagram Group's *The Field Guide to Geology* (New York: Facts on File, 1988), p. 106.

2. *Pedologists* (soil scientists) divide soil into different types. Six basic types are tundra, desert, chernozem, ferralsol, brown forest, and red-yellow podzol. What types of plants grow in each soil type? How does climate affect soil types? What is the composition of the different soil types? For information about soil types, see *The Field Guide to Geology*, pp. 108–109.

3. Soils differ in the size of the particles they contain. Soil types in order of decreasing particle size are sand, silt, and clay. Most soils are not pure sand, silt, or clay, but mixtures of all the types. Such mixtures are called *loam*. How much sand is needed for the soil to be called sandy loam? What is a heavy soil? a light soil? See "soil" in various encyclopedias for information about soil.

Notes on
Soil Texture

Key Facts:

_____

_____

_____

_____

My Results:

_____

_____

_____

_____

_____

_____

_____

_____

_____

Conclusions:

_____

_____

_____

_____

_____

Results of Try New Approaches:

_____

_____

_____

_____

_____

Notes on Designing My Own Experiment:

_____
_____
_____
_____
_____
_____
_____
_____
_____
_____
_____
_____
_____
_____
_____
_____
_____
_____
_____
_____
_____
_____
_____
_____
_____

Notes on Get the Facts:

_____
_____
_____
_____
_____
_____

Crustal Bending
Deformation of the Earth's Crust

Stress acting on rock layers can cause deformation. The results of the past up-and-down and in-and-out movements of the layers are not always apparent from the surface because surface evidence may have worn away over time. Thus, the underlying patterns of deformed layers are often evident only when sections of the Earth are cut away, as with the making of roadways.

In this project, you will demonstrate three types of stress that cause rock deformation—compression, tension, and shear—as well as the different types of deformations that result from each type of stress.

Getting Started

Purpose: To model the formation of an anticline.

Materials

permanent marker   sponge
tap water

Procedure

1. Use the marker to make a line around the perimeter of the sponge through the center of its outside edge.

2. Moisten the sponge with water to make it pliable, then lay it on a table.

3. Without lifting the sponge, place your hands on its short ends and push the ends toward the center of the sponge (see Figure 38.1). Observe the movement and shape of the sponge.

Results

The center of the sponge bends upward in an arch shape.

Why?

The line drawn on the sponge divides the sponge into layers representing **strata** (layers of rock material) in

Figure 38.1

the Earth's crust. The force applied to the sponge represents a form of **stress,** which is a force that acts on rocks in the Earth's crust, causing movement or a change in shape or volume. The type of stress represented in this experiment is **compression** (squeezing together) of rock. Compression can cause rock to break or bend. The movement of the sponge demonstrated a **folding,** or bending of rock layers. A fold producing an upward arch shape is called an **anticline.**

Try New Approaches

A **syncline** is a fold that curves down, creating a troughlike shape. Hold the sponge from the experiment and apply a compression force to cause it to fold downward. By tilting your hands a little, you should be able to first form an anticline, then a syncline.

Design Your Own Experiment

1a. Anticlines are not always visible at the surface. They can be eroded or covered with other materials so that the surface is flat instead of bulging upward. A model of a square cut from the Earth can be made to show the folding of the strata

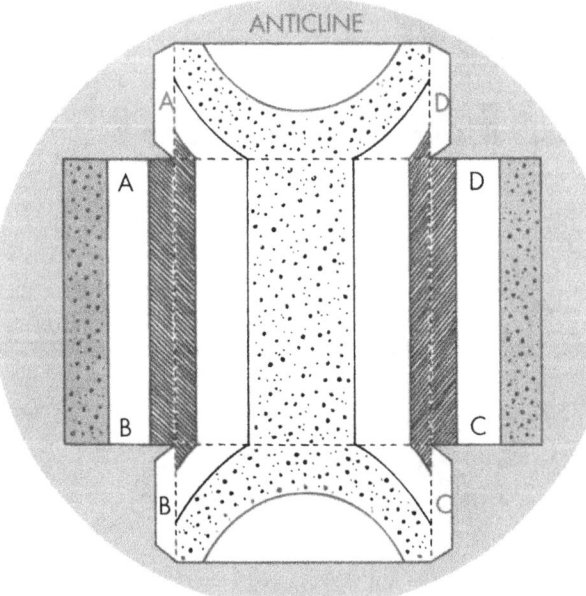

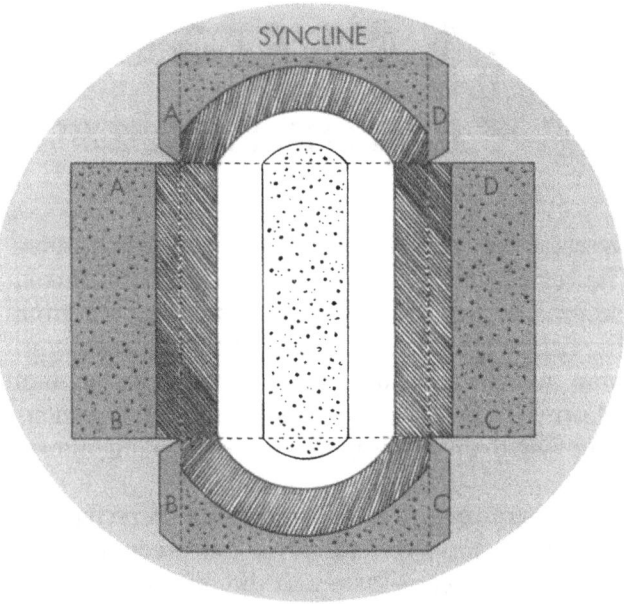

Figure 38.3

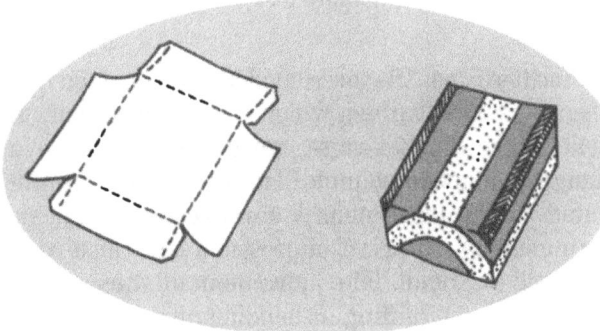

Figure 38.2

beneath the flat surface. Draw a design, such as the one shown in Figure 38.2, on a sheet of typing paper, and color each stratum to indicate different kinds of rocks. (Don't label the tabs or sides.) Cut the diagram out of the paper. Fold the paper along the dashed lines, making all folds in the same direction. Fold the sides over their corresponding tabs—side A over tab A, size B over tab B, and so on. Use tape to secure the tabs to the sides. When standing on its open side, the box will represent an anticline.

b. Prepare a syncline model with a flat surface using a design such as the one shown in Figure 38.3 and the procedure in the previous experiment. Display the two models with labels.

Get the Facts

1. A rock placed under increasing stress goes through three stages of deformation in succession: elastic deformation, ductile deformation, and fracture. What is an elastic limit? Which deformations are irreversible changes? For information about the stages of deformation, see Brian J. Skinner and Stephen C. Porter, *The Dynamic Earth* (New York: Wiley, 1995), pp. 410–411.

2. The Himalayas are the biggest fold mountains on Earth. They are also the largest mountains and have the twenty-eight tallest peaks. What are the characteristics of fold mountains? How were the Himalayas and other fold mountains formed? At what rate are the Himalayas growing? What are other examples of fold mountains? Where are fold mountains generally found? Prepare a display map showing the locations and names of fold mountains. For information about fold mountains, see Steve and Jane Parker, *Mountains and Valleys* (San Diego: Thunder Bay Press, 1996), pp. 20–21.

Notes on
Crustal Bending

Key Facts:

_____
_____
_____
_____

My Results:

_____
_____
_____
_____
_____
_____
_____
_____
_____

Conclusions:

_____
_____
_____
_____
_____

Results of Try New Approaches:

_____
_____
_____
_____
_____

Notes on Designing My Own Experiment:

_____
_____
_____
_____
_____
_____
_____
_____
_____
_____
_____
_____
_____
_____
_____
_____
_____
_____
_____
_____
_____
_____
_____
_____
_____
_____

Notes on Get the Facts:

_____
_____
_____
_____
_____
_____

39 Faulting
The Earth's Crustal Breaking Point

Faulting results when the Earth's crust not only breaks but moves. Stress produces motion in different directions and causes the separate pieces to move in relation to each other. The motion of the crust after breaking is used to classify faults.

In this project, you will study and model the different types of faults and the types of stress that cause them. You will also use models to show the Earth's surface features as a result of faulting.

Getting Started

Purpose: To determine the distinguishing characteristics of a normal fault.

Materials

two lemon-size pieces of clay of different colors

two round toothpicks

table knife

Procedure

1. Break each piece of clay in half.

2. Shape each piece of clay into a roll about 4 inches (10 cm) long.

3. Lay the clay rolls together, one on top of the other, alternating the colors.

4. Press the rolls together into one large clay piece. Flatten the sides of the clay piece by tapping them against a hard surface, such as a table.

5. Use the table knife to cut the clay piece into two parts diagonally.

6. Secure the layers in each part together by inserting a toothpick through the layers, top to bottom.

7. Hold the parts together so that the colored layers match up, then move the left part up and the right part down, as shown in Figure 39.1.

Results

The clay is cut and shifted so that the layers of colored clay in the two parts no longer form continuous horizontal lines.

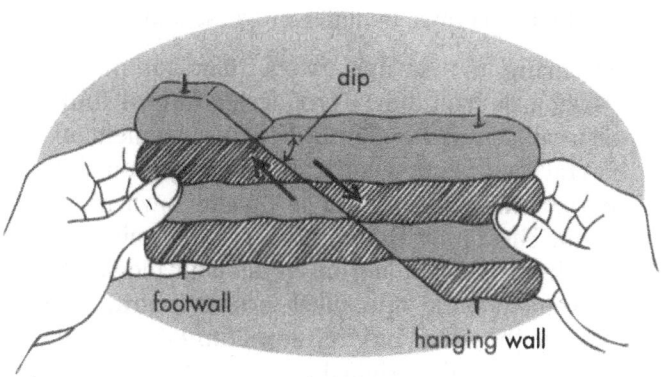

NORMAL FAULT

Figure 39.1

Why?

Each clay color represents a stratum of one kind of rock material. Cutting the clay represents the stress that causes the Earth's crust to fracture. If there is no movement along the fracture, the fracture is called a **joint.** But if there is movement, the fracture is called a **fault.** The fracture line of a fault is called the **fault plane.** If the fault plane shows vertical displacement (up-and-down movement), the **fault block** (rock that bounds a fault plane) above the fault plane is called the **hanging wall** and the fault block below the fault plane is called the **footwall.** In this experiment, the hanging wall of the fault moves down in relation to the footwall. The stress represented is **tension** (the stretching, or pulling apart, of rocks), and the type of fault modeled is a **normal fault.** The angle formed by the fault plane and the top of the hanging wall, measured from the horizontal, is called the **dip** (see Figure 39.1).

While the hanging wall and footwall of the model both moved, it is not always possible to determine whether two fault blocks of the Earth move or whether one stands still while the other moves past it. Fault types are classified by relative displacement, that is, how one side of a fault is pushed out of place in a given direction relative to the other side. With a normal fault, the hanging wall moves down in relation to the footwall.

1. Compression causes a **reverse fault.** This fault is similar to a normal fault, except the hanging wall moves upward in relation to the footwall. Repeat the experiment moving the hanging wall up and the footwall down to model a reverse fault.

2. **Shearing** (stress that twists, tears, or pushes rocks past each other) produces a **lateral fault,** also called a **strike-slip fault.** The movement of a lateral fault along a vertical fault plane is mainly horizontal, with little or no up-and-down movement. The left or right direction is determined by an observer standing on either fault block: the movement of the other block is a **left lateral fault** if it is to the left or a **right lateral fault** if it is to the right. Repeat the original experiment twice, making two clay models. Use one model to represent a right lateral fault and the other a left lateral fault. **Science Fair Hint:** Display the models for each type of fault. Display before-and-after photographs of the position of the clay model representing the different types of faulting.

Design Your Own Experiment

1. A fault block displaced downward and bounded by parallel normal faults is called a **graben** or **rift.** The steep-walled **rift valley** (long, narrow breaks in the Earth's crust) that runs down the center of the mid-Atlantic Ridge is a graben. (For more information about rift valleys in this and other midocean ridges, see Chapter 40, "Plate Tectonics.") An upthrust block bounded by parallel faults is called a **horst.** Use salt dough to make models showing these fault blocks. For each color of dough, mix 2 cups (500 ml) of flour and ½ cup (125 ml) of table salt in a bowl. Add 20 drops of food coloring to ¾ cup (188 ml) of water. Add the colored water to the salt and flour mixture. Knead the dough about 3 minutes or until it is soft and pliable. ***Note:*** Add a little more flour if the dough feels sticky or a little more water if it feels dry.

    Shape three or more 1 × 1 × 4-inch (2.5 × 2.5 × 10-cm) alternating layers of colored dough on a cookie sheet. Cut two diagonal fault planes for

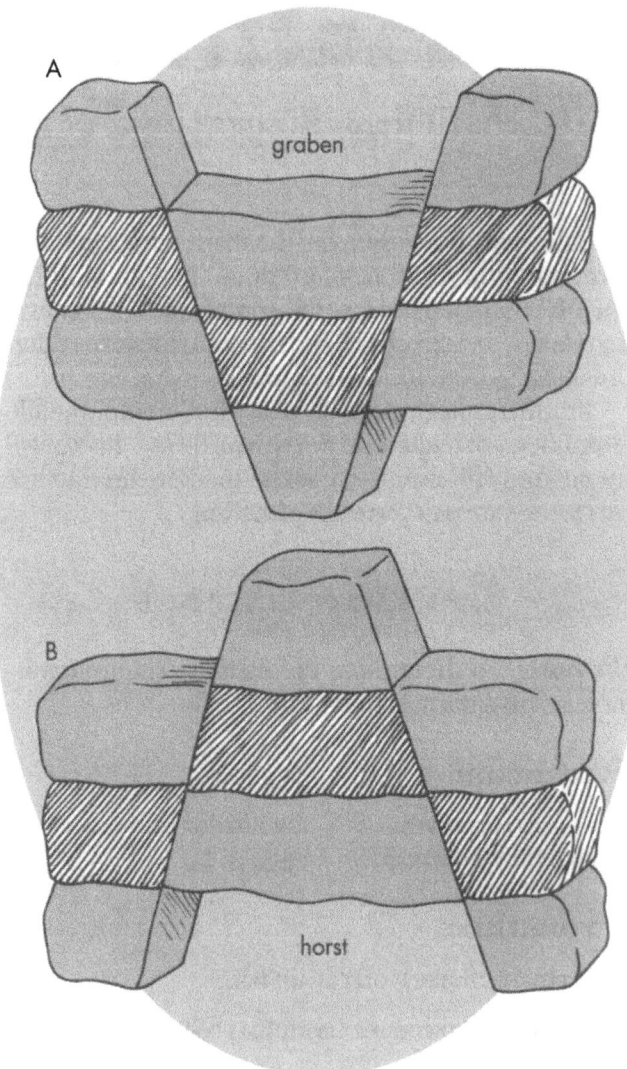

Figure 39.2 A & B

each model. The bottom of the first planes should slant inward with the fault block moved down to represent a graben (see Figure 39.2A). In the second model, the bottom of the planes should slant outward with the fault block moved up to represent a horst (see Figure 39.2B). With adult permission, bake the dough at 275°F (135°C) for 2 hours or until the dough is firm. For information about fault blocks, see David Lambert and the Diagram Group, *The Field Guide to Geology* (New York: Facts on File, 1988), p. 91.

2a. How does tensional stress affect the width of the crust in a **fault zone** (area of Earth's crust that includes the fault blocks on both sides of the fault plane)? On a 4 × 6-inch (10 × 15-cm) piece of cardboard, draw and label the shapes shown in Figure 39.3, then cut out the shapes. Lay the pieces together on a table so that all their edges are even. Measure and record the total width of the assembled pieces. Demonstrate normal faulting caused by tensional stress by moving the two hanging walls down about 1 inch (2.5 cm). Measure and record the total width of the assembled pieces in this normal faulting position.

b. How does compressional stress affect the width of the crust in a fault zone? Assemble the pieces from the previous experiment to form a 4 × 6-inch (10 × 15-cm) rectangle. Demonstrate reverse faulting caused by compressional stress by moving the two footwalls down about 1 inch (2.5 cm), and push the pieces together. Measure and record the total width of the assembled pieces in this reverse faulting position.

c. How would the dip of the fault plane affect the results of the two previous experiments? Repeat each experiment twice. First, use cardboard pieces with a smaller fault plane dip. Then use cardboard pieces with a larger fault plane dip.

Get the Facts

1. The largest recorded abrupt vertical displacement occurred in 1899 at Yakutat Bay, Alaska. Part of the Alaskan shore was lifted as much as 50 feet (15 m) above sea level. Is movement along fault planes always abrupt? How does the depth of the San Andreas fault affect its movement? For information about movement along fault planes, see Brian J. Skinner and Stephen C. Porter, *The Dynamic Earth* (New York: Wiley, 1995), pp. 414–415.

2. The Grand Tetons of Wyoming are fault block mountains. Find out more about the formation of fault block mountains. What is a thrust fault? See Steve and Jane Parker, *Mountains and Valleys* (San Diego: Thunder Bay Press, 1996), pp. 22–23.

3. Faults that are temporarily locked together are called lock faults. How can these and other faults produce earthquakes? For information, see *Janice VanCleave's Earthquakes* (New York: Wiley, 1993), pp. 12–15 and 24–27.

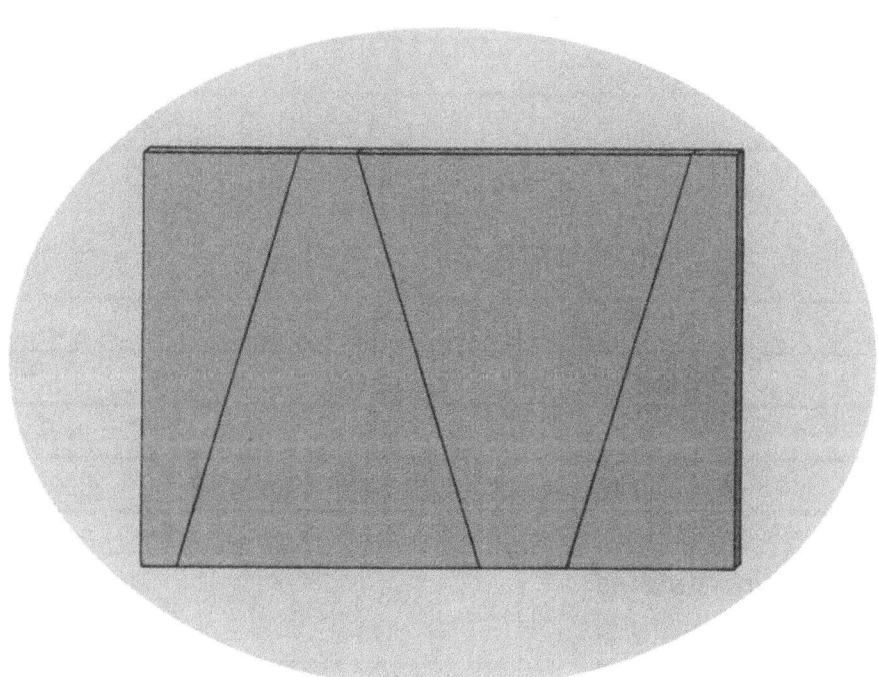

Figure 39.3

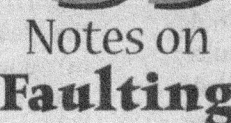

Notes on
Faulting

Key Facts:

_____
_____
_____
_____

My Results:

_____
_____
_____
_____
_____
_____
_____
_____

Conclusions:

_____
_____
_____
_____
_____

Results of Try New Approaches:

_____
_____
_____
_____
_____

Notes on Designing My Own Experiment:

_____
_____
_____
_____
_____
_____
_____
_____
_____
_____
_____
_____
_____
_____
_____
_____
_____
_____
_____
_____
_____
_____
_____
_____
_____

Notes on Get the Facts:

_____
_____
_____
_____
_____
_____

40 Plate Tectonics
Floating Crustal Sections

The Earth's lithosphere is made up of sections called plates that move in relation to each other. This movement is made possible because the plates float on the asthenosphere, which has a thick mudlike texture.

In this project, you will demonstrate seafloor spreading at midocean ridges. You will use models of seafloor spreading to indicate the changes in the Earth's magnetic field over long periods of time. You will learn about the theory of plate tectonics. You will also model the movement of lithospheric plates at divergent boundaries, convergent boundaries, and transform boundaries.

Getting Started

Purpose: To demonstrate seafloor spreading.

Materials

sheet of typing paper
scissors
42-ounce (1.19-kg) empty, round oatmeal box
serrated knife (use with adult approval)

Procedure

1. Fold the paper in half lengthwise with the long edges together.

2. Unfold the paper and cut it in half along the fold line.

3. Use the knife to cut a ¼ × 3 5-inch (0.63 × 12.5-cm) slit in the side of the box.

4. Put the paper strips together, one on top of the other, then push the papers down through the slit in the box. Keep about 2 inches (5 cm) of the strips on the outside and fold them back on opposite sides of the slit.

5. Hold the ends of the strips, one in each hand, and slowly pull about 6 inches (15 cm) of the papers in opposite directions along the surface of the box (see Figure 40.1).

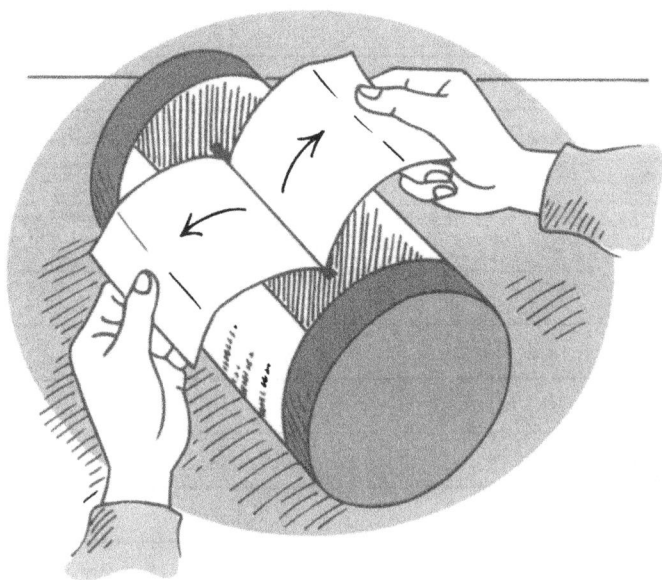

Figure 40.1

Results

The paper strips emerge from the box and move along the box's surface in opposite directions.

Why?

Where the papers exit, the box represents a **midocean ridge** (one of a number of ridges forming a continuous chain of underwater mountains around the Earth). In the center of the midocean ridge is a rift valley. A rift valley is a deep, narrow crack in the Earth's crust, like the slit in the box. Molten rock rises to the surface through this crack. About half of the lava rising out of the rift valley spreads on either side of the midocean ridge. The portion of crust on either side of the ridge is moved apart by the addition of the new material. The lava hardens and forms new ocean floor. This process of the creation of new oceanic crust that moves slowly away from midocean ridges is called **seafloor spreading.**

Evidence of seafloor spreading is a pattern of parallel magnetic "stripes" that are identical on each side of a midocean ridge. The magnetic stripes came about because, before the lava solidified into rock, the mineral grains of magnetic iron in the lava aligned in the direction of the Earth's magnetic field (region in which magnetic materials are acted on by magnetic forces). When the lava solidified, the grains of the rock were permanently fixed in the direction of the Earth's **magnetic field.** But because the Earth's magnetic field has reversed itself many times over millions of years, stripes of rock next to each other may have grains aligned in different directions.

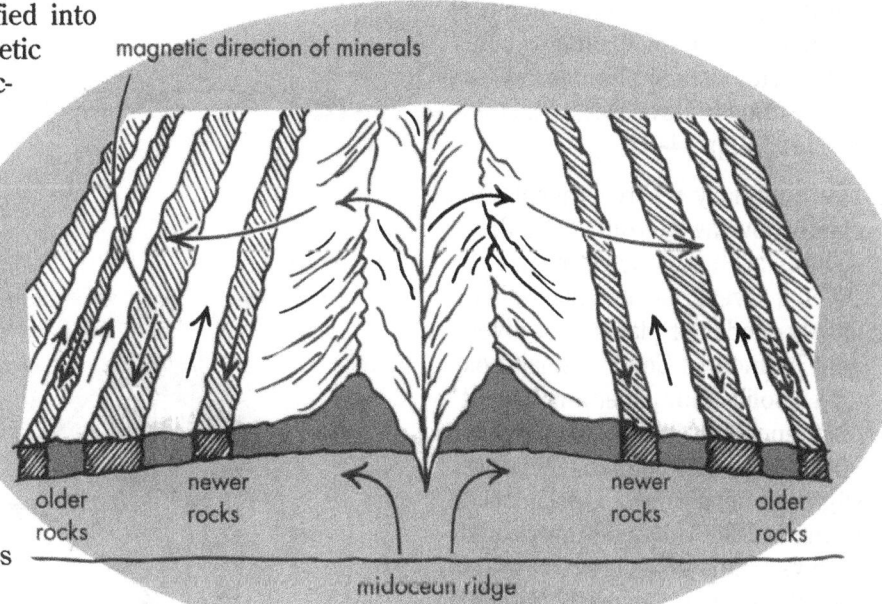

Figure 40.2

Try New Approaches

1. Represent the formation of magnetic stripes on the crust beneath the ocean by pulling the paper strips out of the box so that about 2 inches (5 cm) of the paper comes out of the slit on both sides. Use a colored marker to make a colored stripe about ½ inch (1.25 cm) wide across the paper strips where they come out of the box. Pull the paper strips so that another ½ inch (1.25 cm) comes out the slit. Use a different-colored marker to color the new white paper above the slit. Continue to pull out ½ inch (1.25 cm) of new paper from the slit, alternating the colors until six to eight matching stripes are made on each side of the paper. **Science Fair Hint:** Display the model of the magnetic stripes along with a diagram, similar to Figure 40.2, showing a midocean ridge with magnetic stripes on either side of the ridge. Use arrows to indicate the reversed directions of the stripes.

2. While the seafloor may spread from 1 to 5 inches (2.5 to 12.5 cm) or more per year, the total amount of crust stays the same. This is because as new crust is being formed at the midocean ridges, old crust is sinking into the asthenosphere, where it melts and is absorbed into the mantle. Represent this movement by repeating the original experiment, but cut three slits, 4 inches (15 cm) apart, in the box. Tape both strips of paper, one on top of

the other, to a pencil, near its point (see Figure 40.3A). Wind all but about 6 inches of the paper around the pencil. Position the box on its side, with the center slit on top and a slit on each side. Place the pencil and paper strips inside the box. Separate the strips, pushing one strip out through each of the side slits. Put the ends of the strips together and push them down through the center slit. From inside the box, pull the papers down as far as possible without pulling the strips off the pencil. Slowly turn the pencil to wrap the paper around it as you observe the movement of the paper on the outside of the box (see Figure 40.3B). For more information about seafloor spreading, see David Lambert and the Diagram Group, *The Field Guide to Geology* (New York: Facts on File, 1988), pp. 40–43.

Design Your Own Experiment

1. According to the theory of **plate tectonics,** the Earth's lithosphere is divided into sections called **plates.** These plates float on top of the asthenosphere much like flat rocks on thick mud. Use an earth science text to find out about the lithospheric plates. Prepare a display diagram showing

the shapes, names, and locations of the plates. For information about plate tectonics, see Thomas R. Watters, *Planets* (New York: Macmillan, 1995), pp. 84–85.

2a. The boundary where lithospheric plates move away from each other, such as at the midocean ridges, is called a **divergent boundary.** Prepare a model of plates at a divergent boundary: Cover a small box with a solid-color paper and label it "Asthenosphere." Lay two sponges on top of the box so they are slightly separated. Label each sponge "Plate" and add directional arrows, indicating movement of the plates in opposite horizontal directions. Make a stand-up sign by folding an index card in half lengthwise and labeling the card "Divergent Boundary."

b. Use the previous method to prepare models for a **transform boundary** (a place where two plates slide horizontally in opposite directions alongside each other) and a **convergent boundary** (a place where two plates collide and usually one plate moves under the other). For more information about the differences between these three boundaries, see Keith Stowe, *Essentials of Ocean Science* (New York: Wiley, 1987), pp. 26–35. Make a model showing the boundaries at which crustal material is created, destroyed, and neither created nor destroyed.

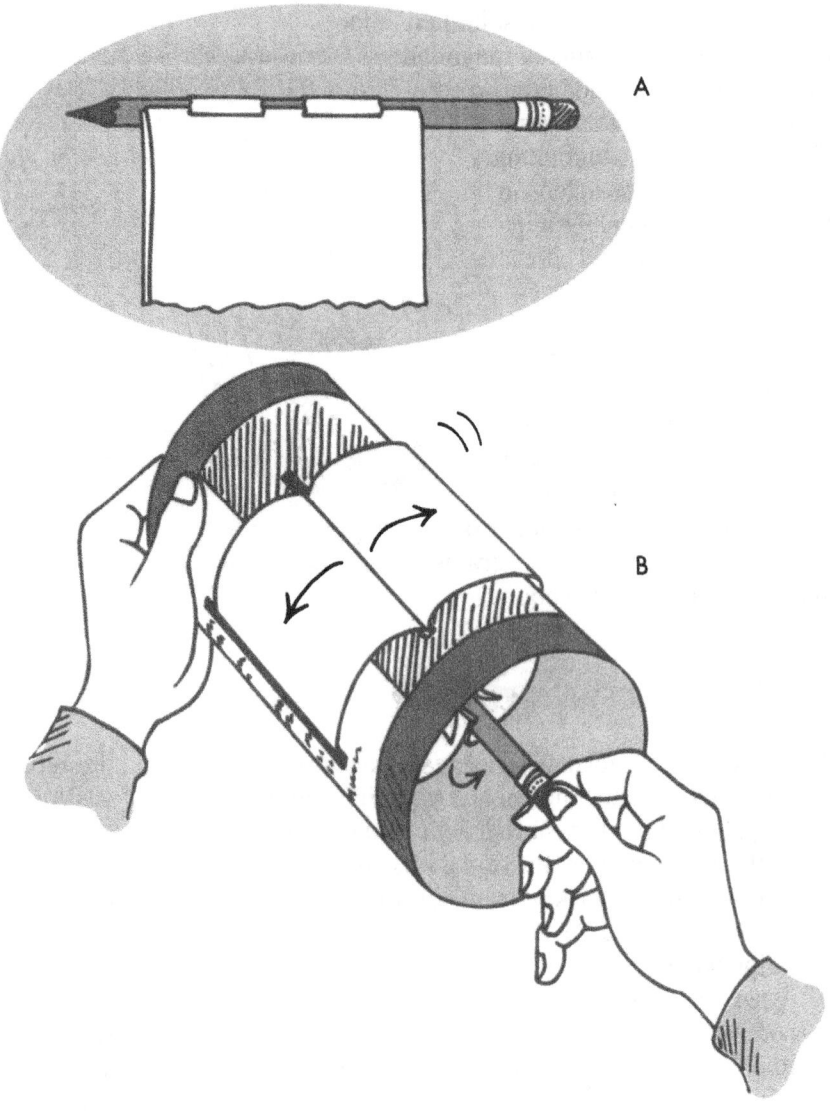

Figure 40.3 A & B

Get the Facts

1. Alfred Wegener (1880–1930), a German scientist, was the first to propose the theory known as continental drift. Find out about Wegener's theory. How is it alike or different from the theory of plate tectonics? What was Pangaea? For information about continental drift, see *Janice VanCleave's Oceans for Every Kid* (New York: Wiley, 1996), pp. 5–11.

2. In 1960, Harry Hess (1906–1969), an American geologist, proposed the theory of seafloor spreading. Find out about the events that led Hess to his conclusion that the seafloor is spreading. For information about Hess, see John S. Dickey, Jr., *On the Rocks* (New York: Wiley, 1988), pp. 145–148.

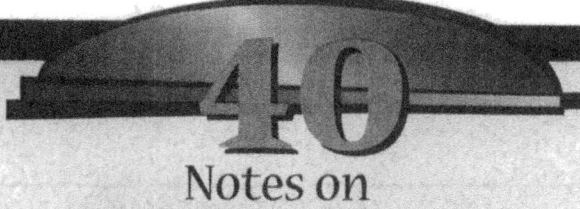

Notes on
Plate Tectonics

Key Facts:

_____
_____
_____
_____

My Results:

_____
_____
_____
_____
_____
_____
_____
_____

Conclusions:

_____
_____
_____
_____
_____

Results of Try New Approaches:

_____
_____
_____
_____
_____

Notes on Designing My Own Experiment:

_____
_____
_____
_____
_____
_____
_____
_____
_____
_____
_____
_____
_____
_____
_____
_____
_____
_____
_____
_____
_____
_____
_____

Notes on Get the Facts:

_____
_____
_____
_____
_____
_____
_____

41 Topography
Highs and Lows of the Earth's Surface

The Earth's surface has features with different elevations, such as mountains, valleys, and lakes, called topography. Three-dimensional (3-D) models, topographic maps, and profile diagrams can be used to show the elevation of surface features.

In this project, you will show how topographic maps use contour lines to indicate the elevation of a land area. You will determine the effect that changes in elevation have on the distance between the map's contour lines. You will use marks called hachures to indicate depressions and craters on your map. You will also measure the elevation of a gently sloping hill, draw a profile map of the hill, and make a 3-D model of the profile of the hill.

Getting Started

Purpose: To produce a 3-D model of a mountain.

Materials

apple-size ball of clay
sheet of copy paper

metric ruler
about 30 toothpicks

Procedure

1. Lay aside a grape-size piece of clay. Use the remaining clay to mold a mountain with an uneven landscape. Make depressions in the side and/or a crater (a hollowed-out area at the top of a volcano) in the top.

2. Set the clay mountain in the center of the paper.

3. Insert the zero end of the metric ruler into the grape-size piece of clay, and stand the ruler vertically next to the clay mountain.

4. On one side of the clay mountain, use a toothpick to draw a straight vertical line from the top of the mountain to its base.

5. Align the vertical line drawn on the clay with the edge of the ruler, then mark heights 1 cm apart on the clay mountain: Holding the toothpick horizontally across the 1-cm mark on the ruler, insert the

end of the toothpick into the line on the clay. Repeat this procedure at each centimeter mark until you reach the top of the mountain (see Figure 41.1).

Figure 41.1

6. Turn the mountain one quarter turn and repeat steps 4 and 5. Repeat the procedure two more times so that heights are marked on four sides of the mountain.

Results

A mountain model with different indicated heights is made.

Why?

In this experiment, you made a 3-D model of a mountain. The toothpicks are placed at different heights to indicate **elevation** (height above ground level or sea level).

Try New Approaches

Topography is the description of the size, shape, and elevation of a region of land. A **topographic map** is a flat map that shows the shapes and **elevations** (height above sea level) of a land area using **contour lines** (lines that connect points on a map that have the same elevation). The difference in elevation between one contour line and the next is called the

contour interval. The contour interval on the map in this experiment is 1 cm.

How can a topographic map of the clay mountain be made? Make the map by removing the ruler and following these steps:

1. Use a pen to trace around the base of the mountain. Make a mark on the paper at each vertical line on the clay model.

2. Wrap an 18-inch (45-cm) piece of dental floss around the mountain at the 1-cm height, letting the floss rest on the toothpicks. Cross the ends of the floss, then pull them in opposite directions to cut through the clay model.

3. Lift the top section of the clay straight up, remove the bottom slice without moving the paper, and lower the top section. The lines on the clay should match the marks on the paper. Trace around the base of the clay.

4. Repeat steps 2 and 3 for the remaining heights marked on the clay with toothpicks.

5. Remove the top, final section of the clay mountain from the paper and observe the tracings on the paper.

Figure 41.2

6. Label the contour lines with the appropriate elevations, as shown in Figure 41.2. Use **hachures** (short lines drawn inside a contour line) to indicate a depression or crater. The free ends of these lines always point downslope.

The closeness of the contour lines indicates the **slope** (the degree of steepness of an inclined surface) of the land. When the lines are far apart, the slope is gentle, but when they are close together, the slope is steep. **Science Fair Hint:** Stack the slices of the clay mountain and display the model along with the topographic map.

Design Your Own Experiment

1a. You can measure the elevation of a small, gently sloping hill by using two rods, such as dowels or plastic pipes, each 2 m long. Using a pen and a meterstick, mark a circle around the center of one rod. Above and below this center mark, use the pen

and the meterstick to mark every tenth of a meter, or decimeter (dm). With 0 at the center mark, first number every decimeter above zero as a positive number (1 to 10), then number every decimeter below zero as a negative number (−1 to −10). Tie a string at least 3 m long around the center of the unmarked rod. Stand the rods together so that the string around the unmarked rod is at the same height as the center mark on the marked rod. Secure the string with tape so that it cannot slide up or down the unmarked rod. Tie the free end of the string loosely to a ring, such as a jar ring, that is larger than the diameter of the marked rod. Place the ring over the marked rod. Stand the rods about 3 m apart and adjust the knot around the ring so the string is taut.

To measure a hill, stand on level ground at the base of the hill with your back to the hill. Hold the unmarked rod upright on the ground. Have your helper stand about 3 m away from you, facing you and the hill. Your helper should hold the marked rod upright so that the negative 10-dm mark is touching the ground and the string between the two rods is taut. Level the string by placing a line level (available in hardware stores) on the string near your rod. Instruct your helper to slowly move the ring up or down the marked rod until you can see the bubble in the center of the line level. This indicates that the string is level. Have your helper note the decimeter mark nearest the ring. The mark should be approximately zero. If the mark isn't zero, both you and your helper should very slowly move away from the hill until the ring at the end of the level string is at the zero mark (see Figure 41.3).

Figure 41.3

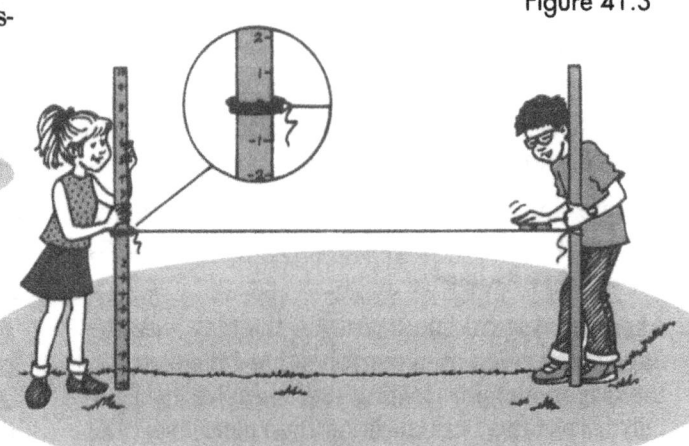

The base of the hill is interval 0, and the measurement is 0 dm. Record this measurement in a Hill Data table similar to Table 26.1. To take the measurement for interval 1, have your helper put his rod in place of yours at the base of the hill. Then move up the hill about 3 m until the string is again taut. Level the string as before, have your helper note the decimeter mark, and record this measurement as interval 1. This and all uphill measurements should be positive numbers. Repeat this procedure, recording each 3-m interval up the hill until you reach the top of the hill. To take downhill measurements, have your helper put his rod in place of yours at the top of the hill. Then move down the hill until the string is again taut. Level the string and record the measurement as before. This and all downhill measurements should be negative numbers. Measure each 3-m interval down the hill until the measurement is again approximately zero and your helper has reached the base of the hill.

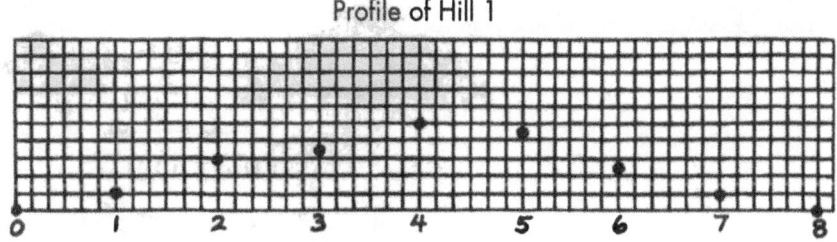

Profile of Hill 1

Figure 41.4

Interval 0, the base of the hill, is marked by a dot labeled 0 in the bottom left corner. Each successive dot is placed six squares, or 3 m, to the right of the preceding dot. The dot is placed above the preceding dot depending on the interval measurement. Thus, because the measurement for interval 1 is 5 dm, the dot for interval 1 is placed one square, or 5 dm, above the dot for interval 0, and so on. The downhill or depression measurements are negative numbers and therefore are placed below the preceding dots. Thus, the dot for interval 5 is 2 cm below the dot for interval 4, and so on.

c. Make a 3-D model of the hill by measuring and graphing the hill at a 90° angle to the first path. Cut out the two profiles from the graph paper and use these as patterns to cut the shapes from stiff paper. Make a vertical cut on each profile so they fit together: Cut one profile halfway down from the highest point; cut the other halfway up from the bottom. Fit the profiles together at a 90° angel (see Figure 41.5).

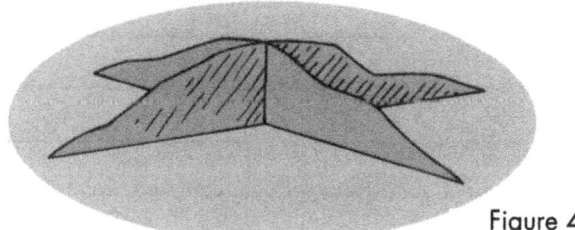

Figure 41.5

TABLE 41.1 HILL DATA	
Interval (3m)	Measurement (dm)
0	0
1	5
2	10
3	2
4	8
5	-2
6	-10
7	-8
8	-5

b. To map the profile (side view) of the hill measured in the previous experiment, follow the procedure described here for the data in Table 41.1 and the graph in Figure 41.4. Each square on the graph paper is equal to 5 dm, and each six squares across represent 30 dm, or 3 m, the measurement interval. The dots labeled 0 to 8 represent the measurements for intervals 0 to 8.

Get the Facts

Cartography is the making or study of maps or charts. Find out more about mapping. What is a relief map? See H. J. de Blij's *The Earth: An Introduction to Its Physical and Human Geography* (New York: Wiley, 1995), pp. 15–26.

41
Notes on
Topography

Key Facts:

_____
_____
_____
_____

My Results:

_____
_____
_____
_____
_____
_____
_____
_____

Conclusions:

_____
_____
_____
_____
_____

Results of Try New Approaches:

_____
_____
_____
_____
_____

Notes on Designing My Own Experiment:

_____
_____
_____
_____
_____
_____
_____
_____
_____
_____
_____
_____
_____
_____
_____
_____
_____
_____
_____
_____
_____
_____
_____
_____
_____

Notes on Get the Facts:

_____
_____
_____
_____
_____
_____
_____

42 The Greenhouse Effect
Heat Transfer in the Atmosphere

The gases in the Earth's atmosphere are warmed by heat radiated from the Earth's surface. These warmed gases surround the Earth and act like a blanket, keeping the Earth warm.

In this project, you will demonstrate the greenhouse effect. You will discover how materials of the Earth's surface affect the greenhouse effect. You will examine the relationship between the greenhouse effect and surface temperatures at night. You will determine how composition and density of the atmosphere affect its ability to trap infrared energy. You will also show how cloud cover affects the surface temperatures at night.

Getting Started

Purpose: To demonstrate the greenhouse effect.

Materials

two shoe boxes	two thermometers
ruler	colorless plastic food wrap
soil	timer

Procedure

1. Cover the bottom of each shoe box with about 2 inches (5 cm) of soil.

2. Lay a thermometer on the surface of the soil in each box.

3. Cover the opening of one box with a single layer of plastic wrap. Leave the other box uncovered.

4. Take readings from both thermometers.

5. Place both boxes side by side in a sunny place outdoors (see Figure 42.1).

Figure 42.1

6. Record readings from both thermometers every 15 minutes for 1 hour.

Results

The temperature readings show that the temperature inside the plastic-covered box was higher and increased faster.

Why?

Radiation, also called **radiant energy,** can move through space, and is not carried by matter. Radiation is also a term for the process by which this energy is transmitted. **Solar energy** is radiant energy from the Sun. Solar energy contains all the different forms of radiant energy, but most of the solar energy reaching Earth consists of **ultraviolet radiation** (causes skin to tan or even burn), **visible light** (radiant energy that can be seen and is divided into the rainbow colors of red, orange, yellow, green, blue, indigo, and violet), and **infrared radiation** (energy in the form of heat given off by all hot bodies).

Radiant energy from the Sun passes through the Earth's atmosphere and reaches the Earth's surface. However, about 30% of the Sun's total radiant energy reaching the Earth is reflected back into space by the atmosphere, the clouds, and the Earth's surface. About 20% is absorbed by the atmosphere, and the remaining 50% is absorbed by the Earth's surface. The radiant energy absorbed by Earth warms its surface, and this warm surface in turn warms the atmosphere above it. An object is warm because of its **internal energy,** which is also called its thermal energy. **Thermal energy** is the sum of all **kinetic energy** (energy possessed by an object because of its motion) and **potential energy** (stored energy of an object due to its condition or position) of particles in random motion making up an object. The faster the particles are moving, the hotter the object. **Heat** is the transfer of thermal energy from one object or region to another due to differences in temperature. The three methods by which heat is transferred from one place to another are: conduction, radiation, and convection. **Conduction** (also called **thermal conduction**) is the method of transferring heat by the collision of one moving molecule with another. Conduction requires physical contact between the bodies or portions of bodies exchanging heat, but radiation does not require contact or the presence of any matter between the bodies. The method of heat transfer called **radiation** is the transfer of heat in the form of infrared radiation. **Convection** is the method of transferring heat by the movement of a heated fluid. Carbon dioxide and water vapor are gases in the atmosphere that help keep heat from being lost to space. They absorb heat from the Earth and then re-emit infrared radiation, some of which strikes and is absorbed by the Earth. Like the plastic covering that prevents the escape of some of the infrared radiation emitted from the soil, as well as the heated **air** (mixture of gases in Earth's atmosphere) inside the closed bowl, the Earth's atmosphere keeps the Earth warm. The absorption of infrared energy by the atmosphere and Earth, called the greenhouse effect, maintains a temperature range on Earth that is hospitable to life. The term **greenhouse effect** comes from the fact that the atmosphere is similar to a greenhouse in that it helps warm the Earth's surface by trapping infrared energy and heated air.

Try New Approaches

1. What effect do surface materials have on the greenhouse effect? Repeat the original experiment, preparing boxes with different surfaces by covering the soil with different materials, such as sand, rocks, and grass. A surface of water could be prepared by lining the box with plastic and filling it with about 2 inches (5 cm) of water instead of soil. **Science Fair Hint:** Display photographs of the various boxes with the results of the experiment. Include a display, such as the one shown in

REFLECTED SOLAR ENERGY

Figure 42.2

Figure 42.2, indicating the percentage of radiant energy reflected back into space. For information about reflected radiant energy, see radiation in Jack Williams's *The Weather Book,* second edition (New York: Vintage Books, 1997).

2. What is the relationship between the greenhouse effect and surface temperatures at night, in the absence of the Sun's radiant energy? Repeat the original experiment, taking temperature readings while the boxes are in direct sunlight outdoors. Then after dark, again take readings from both thermometers every 15 minutes for 1 hour.

3a. Could the composition of the atmosphere affect its ability to trap infrared energy? Compare different materials for their ability to trap infrared energy. Repeat the original experiment using plastic wrap and other materials, such as waxed paper, clear Plexiglas, and glass. **Science Fair Hint:** Display samples of the box covers with the results of the experiment.

b. **Density** is the mass of a substance per unit volume. The greater the density of a substance, the closer together its particles. How does the density of the atmosphere affect its ability to trap infrared energy? Repeat the previous experiment twice, first using two layers of covering material, then using three layers. **Science Fair Hint:** Compare the results of the experiment to surface environments on celestial bodies with little or no atmosphere, such as the Moon and Mars, and those with a dense atmosphere, such as Venus. Use an astronomy text to find out about the atmosphere of the different celestial bodies. Make charts showing the composition and density of their atmosphere and surface environment.

Design Your Own Experiment

1. Design a way to determine how the Earth's atmosphere affects the surface air temperature at night. One way is to record air temperature at sunset and again at sunrise for one or more weeks in a Day and Night Temperature Data table similar to Table 42.1. Calculate the difference between the two temperatures each day, and determine an average by adding the differences and dividing by the total number of days. *Note:* Your temperature meas-

TABLE 42.1 DAY AND NIGHT TEMPERATURE DATA

Day	Sunset Temperature, °F (°C)	Sunrise Temperature, °F (°C)	Temperature Difference, °F (°C)
1	80° (26.7°)	62° (16.7°)	18° (10°)
2	82° (27.8°)	61° (16.1°)	21° (11.7°)
↕			
7	80° (26.7°)	63° (17.2°)	17° (9.5°)

urements should be taken during a time of a constant weather pattern.

2. Repeat the previous experiment during a one-month time span to determine how cloud cover affects surface temperature during the night. Do this by recording air temperature during cloudy and noncloudy periods. On four or more days on which the weather forecast calls for the same amount of nighttime cloud cover, record the temperature at sunset and again at sunrise. Repeat by recording the day and night temperatures on four or more days with little or no nighttime cloud cover. Use the results to determine how the presence or absence of clouds in the troposphere leads to more heat escaping to space, thus causing a greater decrease in nighttime temperature.

Get the Facts

Carbon dioxide, one of the greenhouse gases, is responsible for much of the warming of the Earth. Some scientists predict a rise in the average temperature of the Earth if the amount of carbon dioxide in the atmosphere continues to increase. Find out more about the production of carbon dioxide. How do fossil fuel emissions and deforestation affect the level of carbon dioxide in the atmosphere? What is insolation and how does it affect global warming? For information about the greenhouse effect and greenhouse gases, see *Janice VanCleave's Ecology for Every Kid* (New York: Wiley, 1996), pp. 139–146.

Notes on
The Greenhouse Effect

Key Facts:

_____

_____

_____

_____

My Results:

_____

_____

_____

_____

_____

_____

_____

Conclusions:

_____

_____

_____

_____

_____

Results of Try New Approaches:

_____

_____

_____

_____

_____

Notes on Designing My Own Experiment:

_____
_____
_____
_____
_____
_____
_____
_____
_____
_____
_____
_____
_____
_____
_____
_____
_____
_____
_____
_____
_____

Notes on Get the Facts:

_____
_____
_____
_____
_____

Barometric Changes
The Cause and Measurement of Air Pressure

Atmospheric pressure is the measure of the pressure that the atmosphere exerts on surfaces. Since the atmosphere is composed of air, the term *air pressure* is sometimes used. Weather forecasters measure atmospheric pressure with barometers and use the term *barometric pressure*.

In this project, you will make different barometers and use them to measure and compare atmospheric pressure in different places and at different times. You will demonstrate how the impact of air molecules exerts atmospheric pressure. You will show how air with greater density exerts greater atmospheric pressure. You will also learn about the relation between barometric pressure at sea level and at higher altitudes as well as natural barometers.

Getting Started

Purpose: To show how a barometer works.

Materials

serrated knife (use with
   adult approval)
scissors
masking tape
two 20-ounce (600-ml)
   plastic bottles
one-hole paper punch
flexible straw
tap water

Procedure

1. Cut off the top 3 inches (7.5 cm) of one of the bottles. Do this by sawing a small slit in the bottle with the knife, then use the scissors to cut around the bottle. Keep the bottom section.

2. Cover the cut edges of the bottle with tape.

3. Use the paper punch to make a hole about 1 inch (2.5 cm) from the cut edge of the bottle.

4. Insert about ½ inch (1.25 cm) of the straw through the hole. The straw should fit snugly in the hole.

5. Stand the cut and uncut bottles side by side. Fill the bottles half full with equal amounts of water.

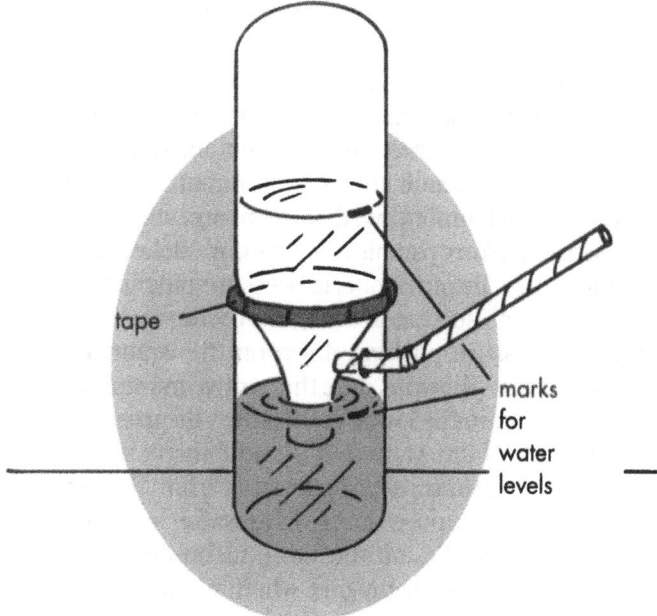

tape

marks
for
water
levels

Figure 43.1

6. With your fingers over the mouth of the uncut bottle, turn it upside down, then lower it into the cut bottle. The mouth of the bottle should be below the water level in the cut bottle, and the two bottles should fit snugly together. The combined bottles form a barometer.

7. Near the straw, use the marking pen to mark the water lines on both bottles (see Figure 43.1).

8. Blow into the straw and observe the change in the water level in each bottle.

9. Suck air out of the straw, and again observe the change in the water level in each bottle.

Results

Blowing into the straw causes the water to rise above the water line in the top bottle and sink below the line in the bottom bottle. The reverse is true when air is sucked out of the straw: the water sinks below the water line in the top bottle and rises above it in the bottom bottle.

Why?

Atmospheric pressure, also called air pressure, is the force that air exerts on a particular area. It is a measure of the pressure resulting from the **force** (push or pull exerted on matter) that the total mass of air in an imaginary column exerts on a horizontal area. The greater the mass—that is, the greater the number of molecules—the greater the air pressure.

Atmospheric pressure is directly related to the density of air, which measures the number of air molecules in a given space. This was demonstrated by the barometer you made in this experiment. A **barometer** is an instrument used to measure atmospheric pressure. Before you blew air into or sucked air out of the straw, the atmospheric pressure pushing down on the water's surface in the lower bottle was equal to the air pressure pushing down on the water in the upper bottle. Blowing into the straw increased the amount of air in the lower bottle; thus the pressure on the surface of the water in this bottle increased. This increase in pressure forced water from the lower bottle into the upper bottle. The water level in the lower bottle sank, and the level in the upper bottle rose. The reverse happened when air was removed from the lower bottle. The pressure pushing down on the water in the lower bottle decreased, and the pressure in the upper bottle pushed water down into the lower bottle. The water level in the upper bottle sank, and the level in the lower bottle rose.

Try New Approaches

1a. Does atmospheric pressure outside the bottle barometer change enough to affect the water levels in the bottles? Make a scale by sticking a 3-inch (7.5-cm) strip of masking tape along the side of and centered over the mark of the top bottle. Use a fine-point pen and a ruler to mark a line in the center of the tape over the mark on the bottle. Continue to mark lines ⅟₁₆ inch (2 mm) apart above and below the center line. Number the center line 0. Number the tenth mark below the line –1 and the next tenth mark –2. Continue to the end of the tape. Repeat, using positive numbers above the center line. Set the barometer inside your home where it will not be disturbed. Leave the straw open. At the same time each day for seven or more days, record the water level on the scale in the top bottle.

b. Does the atmospheric pressure outside your house differ from the pressure inside? Repeat the original experiment, making a second bottle barometer. Then, repeat the previous experiment, placing one barometer inside and the other outside. **Science Fair Hint:** Record the atmospheric pressure from a local television or radio weather program. Prepare a diagram showing the bottle barometer's readings and the recorded atmospheric pressure for each day.

2. Does atmospheric pressure change during the day? Use the bottle barometer to record the water level and time every hour for 6 or more hours during one day. Use the information to prepare a bar graph showing the pattern of change.

Design Your Own Experiment

1. Prepare another type of barometer by cutting the top from a 12-inch (30-cm) round balloon. Stretch the bottom section of the balloon across the top of a wide-mouthed 1-quart (1-liter) jar, and secure it with a rubber band. Cut one end of a drinking straw to a point. Glue the uncut end of the straw to the center of the stretched balloon. Secure a metric ruler to another 1-quart (1-liter) jar with a rubber band. The ruler must stand upright, with the zero end of the metric scale at the bottom. (The marks on the metric scale are close, allowing small pressure changes to be measured.) Position the two jars so that the pointed end of the straw points to the metric markings on the ruler. Do not allow the straw to touch the ruler (see Figure 43.2). Repeat the previous experiments replacing the bottle barometer with this jar barometer.

2a. Air is made up of gas molecules, mostly nitrogen and oxygen. These molecules move and hit against each other and anything that gets in their path. The impacts of these bouncing molecules cause pressure. Demonstrate how the impact of an air molecule causes atmospheric pressure by dropping a marble on a scale. Use a scale with a 1-pound (454-g) capacity. Place the scale in the center of a box that is about 1 foot (30 cm) long and wide and about 6 inches (15 cm) taller than the scale. Hold a marble about 6 inches (15 cm)

Figure 43.2

above the scale and drop it. (The box just needs to be large enough to contain the marble when it falls off the scale.) Observe the movement of the scale's dial.

b. Air molecules are in constant motion. Like air molecules hitting against a surface, enough marbles hitting the scale in succession would keep the scale indicator from moving back to zero. Design a method to drop a series of marbles, such as rolling them down a folded piece of cardboard toward the scale.

Get the Facts

1. Evangelista Torricelli (1608–1647), an Italian mathematician and physicist, discovered the principle of a barometer in 1643. An encyclopedia can be used to find out about the experiment that Tor-

ricelli did to discover how the pressures of air could be measured. A water barometer was invented by Otto von Guericke (1602–1686), the mayor of Magdeburg, Germany, in 1646. How did Guericke's barometer compare to the barometer designed by Torricelli?

2. Aneroid barometers are inexpensive and commonly found in homes and offices. Aneroid is from a Greek word meaning "without liquid." Find out how this barometer measures air pressure. For information about aneroid barometers, see Frank H. Forrester, *1001 Questions Answered about the Weather* (New York: Dover Publications, 1981), p. 17.

3. Barometric pressure increases at altitudes above sea level. What causes this difference in pressure due to altitude? For information, see *The Nature Company Guides: Weather* (Time Life Books, 1996), p. 26.

NAME

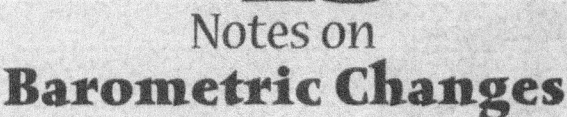

Notes on
Barometric Changes

Key Facts:

_____

_____

_____

_____

My Results:

_____

_____

_____

_____

_____

_____

_____

Conclusions:

_____

_____

_____

_____

_____

Results of Try New Approaches:

_____

_____

_____

_____

_____

Notes on Designing My Own Experiment:

_____
_____
_____
_____
_____
_____
_____
_____
_____
_____
_____
_____
_____
_____
_____
_____
_____
_____
_____
_____
_____
_____

Notes on Get the Facts:

_____
_____
_____
_____
_____
_____

44. Soundings
Mapping a Profile of the Ocean Floor

The first scientific attempt to measure the depth of the ocean was made during an expedition by the British ship H.M.S. *Challenger* between 1872 and 1876. The measuring method on the *Challenger* was called sounding and involved lowering weighted ropes to the ocean bottom. Today investigators use echo-sounding sonar and other methods to determine ocean depths.

In this project, you will model the use of sonar to determine ocean depths. You will also determine how to use echo soundings to graph a profile of the ocean floor and learn how distances between echo soundings affect accuracies of profiles.

Getting Started

Purpose: To model the use of sonar to determine ocean depth.

Materials

tennis ball
helper

stopwatch
calculator

Procedure

1. Hold your arms against the side of your body, then bend your elbows at a 90° angle so that your hands are held straight out in front of you.

2. Hold the ball in one hand.

3. Standing still and keeping your elbows against your sides, practice bouncing the ball several times until you can throw it with the least amount of force necessary to cause it to return to your other hand (see Figure 44.1).

4. Ask a helper to measure the time it takes for the ball to leave one hand and return to the other. When your helper says "go" and starts the stop-

watch, throw the ball. When you catch the ball, say "stop" and have your helper stop the watch and record the time that has passed as the echo time.

5. Repeat step 4 three more times and average the results by adding together the four recorded echo times and dividing the sum by 4.

   Example:

   0.56 sec + 1.02 sec + 0.59 sec +1.0 sec = 3.17 seconds

   3.17 sec ÷ 4 = 0.7925 = 0.79 seconds

6. The depth of the ocean floor (*D*) is one-half the average echo time (*t*) multiplied by the speed of sound in water (*S*). Use the average echo time to model how the depth in water could be determined by the following formula:

$$D = \tfrac{1}{2}t \times S$$

The speed of sound in ocean water is about 5,000 feet per second (1,500 meters per second), which can be written 5,000 ft/sec (1,500 m/sec).

Example:

$D = \tfrac{1}{2} \times 0.79$ sec $\times 5,000$ ft/sec (1,500 m/sec)

= 1,975 feet (592.5 m)

Results

The time for the ball to bounce and return will vary depending on the height of the person throwing it. For the example, the average time was 0.79 seconds. Using this as a model of the sonar echo time in the ocean, the ocean depth was calculated to be 1,975 feet (592.5 m).

Why?

Mapping the ocean floor requires methods different from those used on land. In the past, the depth was measured by a method called **sounding.** Sounding has nothing to do with sound. To take a sounding, knots were tied in a rope at intervals of 1 **fathom**

Figure 44.1

(6 feet, or 1.8 m). A weight was tied at one end of the rope and dropped over the side of the ship. The number of knots that went over the side before the weight struck bottom was counted. The number of knots equaled the depth in fathoms.

A modern method of measuring ocean depth is **echo sounding,** which is a method of sending out sound from a transmitter and measuring the **echo time** (time it takes sound leaving a transmitter to travel to an object, be reflected, and return to a receiver). Echo sounding is often called sonar. The term **sonar** is an acronym derived from *SO*und *Na*vigation *A*nd *R*anging. Sonar is the method or the device used to determine ocean depth or distance by calculating the echo time of sound.

In this experiment, the bouncing ball represents sound bouncing off the ocean floor. The time for the round trip—the echo time—was measured and used to calculate the one-way distance—the depth.

Try New Approaches

1a. Design a model for measuring the depth of different parts of the ocean floor by taping a 4-foot (120-cm) piece of adding machine tape to a tile floor. Write 0 at one end and mark each foot (30 cm) from 1 foot to 4 feet (30 cm to 120 cm), as shown. Position a stool on the zero mark and a chair at the 1-foot (30-cm) mark, and stack five or six thick books at the 2-foot (60-cm) mark and two or three books at the 4-foot (120 cm) mark (see Figure 44.2). Now, using the stool, chair, and the books as part of the ocean floor, repeat the original experiment, measuring the echo sound by bouncing the ball at each of the four marks. At zero, the stool that represents the shoreline, record the time as zero, then move to the 1 foot (30-cm) mark, drop the ball, and time the echo. Continue moving to each mark on the

tape, dropping and timing the bouncing ball. Record the measurements in an Ocean Depth Data table like Table 44.1. Use the depth equation to calculate the depth at each distance from the shoreline, and record it in the data table.

TABLE 44.1 OCEAN DEPTH DATA		
Distance from Shoreline	**Average Echo Time**	**Calculated Depth**
0 foot	0 second	0 foot (0 m)
1 foot (30 cm)	0.35 second	875 feet (262.5 m)
2 feet (60 cm)	0.52 second	1,300 feet (390 m)
3 feet (90 cm)	0.80 second	2,000 feet (600 m)
4 feet (120 cm)	0.68 second	1,700 feet (510 m)

b. Use the calculated depths from the data table to plot a graph representing the profile of the ocean floor (see Figure 44.3).

OCEAN MODEL PROFILE

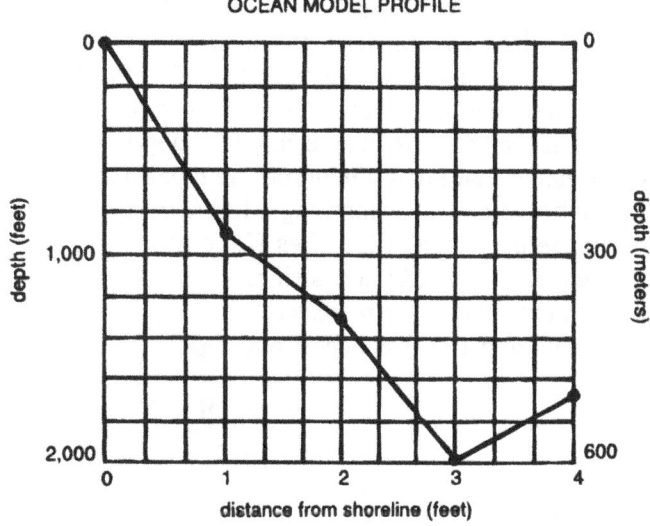

Figure 44.3

c. How does the distance between the echo soundings affect the accuracy of the profile? Repeat parts 1a and 1b taking echo soundings every 6 inches from the model's shoreline. Compare the profiles made with 1-foot (30-cm) soundings and 6-inch (15-cm) soundings to determine which better represents a profile of the ocean model.

Figure 44.2

Design Your Own Experiment

1a. Design another ocean model representing a section of the ocean, and use sounding to map the ocean model's profile. One way is to place two identical chairs, with backs at least 30 inches (70 cm) tall, 4 feet (1.2 m) apart. Represent the surface of the ocean by tying a string horizontally between the highest points of the chairs. Use a black marking pen to mark off 3-inch (7.5-cm) intervals along the "surface" string. Place stacked books, a stool, and an upturned pot and bowl under the string as shown in Figure 44.4. These objects represent features on the ocean floor.

Cut a second string 12 inches (30 cm) longer than the height of the chairs. Tie a washer to one end. Use the pen to mark off a 1-inch (2.5 cm) scale along this second string. This string will be called the scale. Holding the free end of the scale, position it against the surface string, next to the back of one chair (this is the 0-inch, or 0-cm, mark) and slowly lower the scale until the washer touches an object or the floor. Use the marks on the scale to determine the depth of the ocean at that point. Round off the measurement to the nearest scale marking. Measure the depth of the ocean every 3 inches (7.5 cm) along the length of the surface string, and record the depth measurements and the distances from the shoreline (the zero mark) in a data table like Table 44.1.

b. Use the data table to make a graph of your measurements. Title the graph "Ocean Model Profile." Display the graph and a photograph of the profile of the ocean model.

Get the Facts

1. The profile of the ocean floor is divided into four areas: the continental shelf, the continental slope, the continental rise, and the abyss. Which areas make up the *continental margin* (water-covered area from the shoreline of the continents to the abyss)? What is the location and size of each area? For information about these four ocean areas, see Don Groves's *The Oceans* (New York: Wiley, 1989), pp. 96–97. Display a diagram of the ocean profile showing the four areas.

2. On December 21, 1872, H.M.S. *Challenger* embarked from Portsmouth, England, and changed the course of scientific history. Physicists, chemists, and biologists collaborated with expert navigators to map the sea. During the 4-year journey, the voyager circumnavigated the globe and sounded the ocean bottom to a depth of 26,850 feet. Find out about this historic journey. What equipment was used to make the soundings? Information can be found by searching for H.M.S. *Challenger* on the Web.

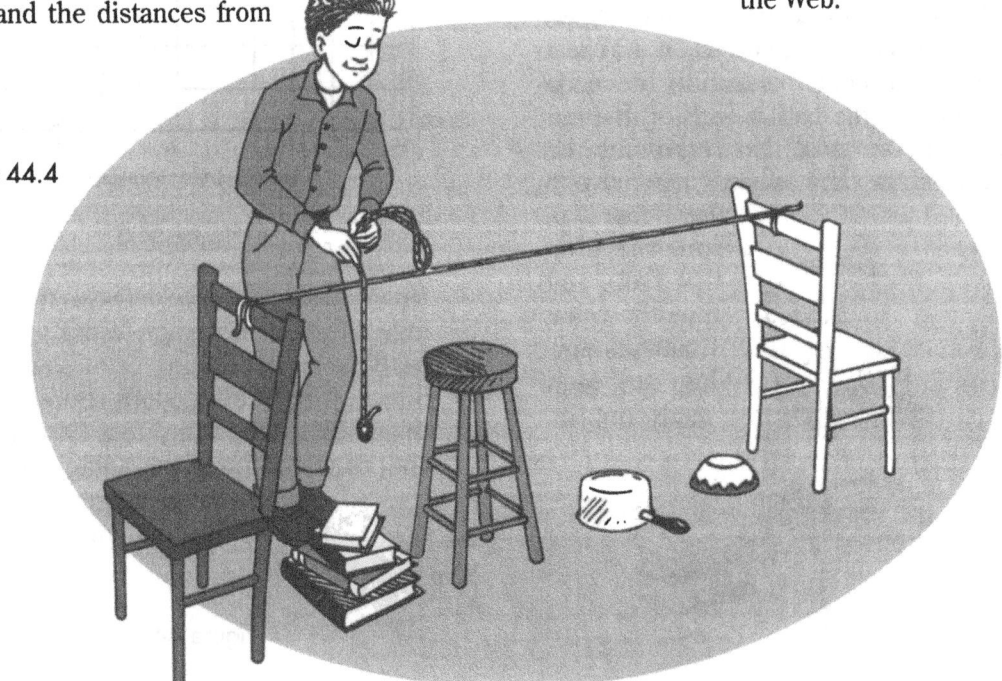

Figure 44.4

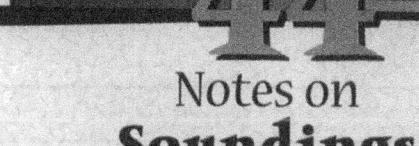

Notes on
Soundings

Key Facts:

_____

_____

_____

_____

My Results:

_____

_____

_____

_____

_____

_____

_____

Conclusions:

_____

_____

_____

_____

_____

Results of Try New Approaches:

_____

_____

_____

_____

_____

Notes on Designing My Own Experiment:

_____
_____
_____
_____
_____
_____
_____
_____
_____
_____
_____
_____
_____
_____
_____
_____
_____
_____
_____
_____
_____
_____
_____
_____

Notes on Get the Facts:

_____
_____
_____
_____
_____
_____
_____

Physics

45 Resonance
Sympathetic Vibration

As soldiers walk across a bridge, they cause the bridge to vibrate. If the soldiers march in rhythm with the natural frequency of the bridge, each step will cause the bridge to vibrate at a higher amplitude. If the amplitude is great enough, the bridge could actually collapse. This phenomenon in which a small, repeated force causes the amplitude of a vibrating object to become very large is called resonance.

In this project, you will demonstrate sympathetic vibration, or resonance, and determine how distance affects sympathetic vibration. You will investigate how building size and stiffness affect the resonance caused by earthquakes. You will also learn how the application of a force at the same natural frequency of an object affects the amplitude of the object's motion.

Getting Started

Purpose: To demonstrate sympathetic vibration.

Materials

2 identical empty 1-liter plastic soda bottles

Procedure

1. Blow across the mouth of one soda bottle to produce a constant sound. Note the pitch and the loudness of the sound.

2. As you blow across the mouth of the bottle, place the mouth of the other bottle near your ear as shown in Figure 45.1. Note any change in the pitch and the loudness of the sound.

Results

When you blew across the mouth of the first bottle alone, you heard a sound. When you blew across the mouth of the first bottle while the second bottle was held next to

Figure 45.1

your ear, you heard a sound that had the same sound but was louder.

Why?

Frequency is a term used in physics to denote the number of times that any regularly recurring event, such as **vibrations** or **oscillations** (swinging or back-and-forth movements) occur in one second. **Resonance** is the condition of starting or amplifying vibrations in a body at its natural frequency by a vibrating force having the same frequency: this is also called **sympathetic vibration.** All objects have a natural frequency depending on their size and shape. Since the two soda bottles are alike in size and shape, they share the same natural frequency. Blowing across the first bottle causes the air inside the bottle to vibrate, which makes the air around the bottle's mouth vibrate as well. This vibrating air moves outward and causes the air in the second bottle to start vibrating. Resonance occurs when the natural frequencies of two objects are the same or if one has a natural frequency that is a multiple of the other. For example, if the natural frequency is 4 vibrations/1 sec, a multiple that is two times the natural frequency would be 8 vibrations/1 sec. The second bottle in this experiment, vibrated even without air being blown across its mouth because the air entering it was vibrating at the same frequency as its natural vibration. Since the rate of vibration for the two bottles was the same, there was **constructive interference,** meaning the vibrations added together producing a vibration with a larger **amplitude** (the farthest displacement of an object from its resting position), and thus a louder sound. The vibrating air makes a sound that you can hear when the vibrations reach your ears. So the combined vibrating air from the two bottles produced a louder sound but at the same **pitch** (how high or how low a sound is).

Try New Approaches

How does the distance of a vibrating source affect resonance? Repeat the experiment asking a helper to hold a bottle near his or her ear while you blow across your bottle's mouth. First repeat the experiment standing about 3 feet (1 m) from your helper, then standing about 6 feet (2 m) away. Reverse positions and ask your helper to blow across the mouth of one of the bottles while you hold the second bottle near your ear.

Figure 45.2

Design Your Own Experiments

1a. An **earthquake** (violent shaking of the Earth caused by the sudden movement of rock beneath its surface) can cause more resonance in some buildings than in others. Is the difference because of the size of buildings? Design a way to show how size affects resonance. One way is to verify that as the circumference of paper rings decreases the frequency at which they strongly vibrate or resonate increases. Prepare the paper strips by cutting two 1-inch (1.25-cm)-wide strips from a sheet of copy paper. One strip should be 10 inches (25 cm) long and the other 8 inches (20

cm) long. Tape the ends of each strip together to form two circular rings. Tape each ring to the center of a 4-by-4-inch (10-by-10-cm) piece of cardboard so that they are about 1 inch (2.5 cm) apart as shown in Figure 45.3. Shake the cardboard from side to side. Start at a low frequency by slowly moving the cardboard back and forth. Then gradually increase the frequency of your shaking by increasing the speed at which it is shaken. Notice when each of the paper rings starts to vibrate.

b. What effect does stiffness of a material have on resonance? Repeat the previous experiment, using stiff paper to make the circles. Attach the stiff circles to the ends of the cardboard used in step 1a so that you have the different types of paper on the same piece of cardboard. Observe the same-size rings while you shake the cardboard and determine if their frequency varies.

2. Resonance also can be induced when a small force is applied at the right frequency. An example of this is pumping a swing. Design an experiment to demonstrate that when you apply a force at intervals close to the natural frequency of an object, the amplitude of the object will increase. Since the frequency of a **pendulum** (a suspended object called a **bob** that swings back and forth) depends on its length, create three or more pendulums from different lengths of string. Stretch and tie a 1-yard (1-m) string between two chairs. Tie a washer (bob) to the ends of three strings of 12 inches (30 cm), 10 inches (25 cm), and 8 inches (20 cm) long. Tie the free ends of these strings to the horizontal string between the

Figure 45.3

chairs as shown in Figure 45.4. At regular intervals, very gently tap the side of the horizontal string, near one of the chairs. Change the frequency of the tapping (increasing or decreasing) until one of the pendulums starts to swing. Once a pendulum starts to swing, continue tapping at this frequency for 20 swings. You can determine the frequency of the pendulum by timing the 20 swings. Then calculate the frequency using this equation: F = vibrations (swings)/time.

For example, if the time for 20 swings is 10 seconds, the frequency would be:

$$F = 20 \text{ vibrations}/10 \text{ sec}$$
$$= 2 \text{ vibrations/sec}$$

3a. Determine how the movement of one pendulum can affect the movement of another pendulum with the same frequency. Tie a fourth pendulum 12 inches (30 cm) long, on the horizontal string from the previous experiment next to the other 12-inch (30-cm) pendulum. Pull one of the 12-inch (30-cm) pendulums toward you and release. Observe the movement of each pendulum.

b. Does the position of pendulums of the same length affect their movement? Repeat the previous experiment twice, using three pendulums, each 12 inches (30 cm) long. First tie the pendulums on the horizontal string 2 inches (5 cm) apart, then repeat by tying the pendulums 6 inches (15 cm) apart.

Get the Facts

1. A natural vibration is also called a *free vibration*. What is a forced vibration? For information see Sir James Jeans's *Science and Music*. (New York: Dover Publication, Inc., 1968), pp. 53–54.

2. How did resonance destroy the Tacoma Narrows Bridge in Washington State on November 7, 1940? For information see P. Erik Gundersen's *The Handy Physics Answer Book* (Detroit: Visible Ink, 1999), pp. 197–198.

Figure 45.4

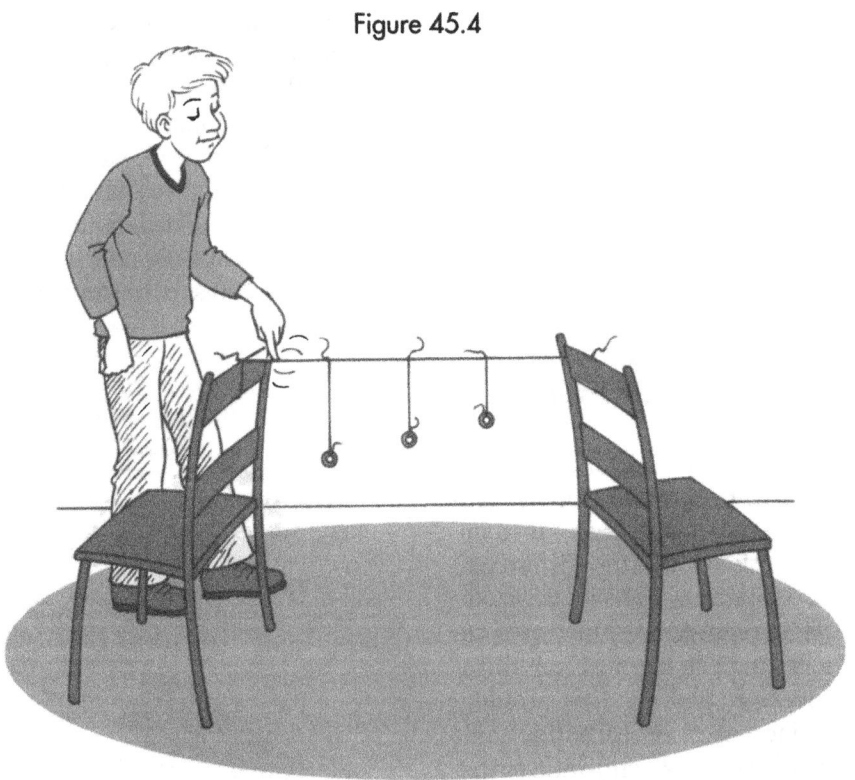

Notes on
Resonance

Key Facts:

_____
_____
_____
_____

My Results:

_____
_____
_____
_____
_____
_____
_____

Conclusions:

_____
_____
_____
_____
_____

Results of Try New Approaches:

_____
_____
_____
_____

Notes on Designing My Own Experiment:

_____
_____
_____
_____
_____
_____
_____
_____
_____
_____
_____
_____
_____
_____
_____
_____
_____
_____
_____
_____
_____
_____

Notes on Get the Facts:

_____
_____
_____
_____
_____
_____

46 Thermal Conduction
Transfer of Vibrational Energy

The term heat is often used very loosely. It is commonly said that hot objects possess more heat than do cold objects but technically the energy in the hot object is not heat, it is thermal energy. Heat is the movement of thermal energy from one object to another as a result of differences in temperature. The process by which heat is transferred from one particle to another due to collisions of the particles is called thermal conduction or conduction.

In this project, you will determine the effects of distance, time, cross-sectional area, and types of materials and temperature on thermal conduction. You will also compare the efficiency of different thermal insulators.

Getting Started

Purpose: To determine how distance affects thermal conduction.

Materials

lemon-size piece of
   modeling clay
one small birthday candle
metal cookie sheet or
   baking pan
1 small paper clip
permanent marker

metric ruler
margarine
18-by-30-inch (45-by-
   75-cm) piece of
   aluminum foil
transparent tape
kitchen matches

Procedure

1. Use a grape-size piece of clay to stand the candle on the metal cookie sheet, as shown in Figure 30.1.

2. Unbend the paper clip to form a metal wire that is as straight as possible.

3. Use the marker and the ruler to make marks on the wire at 2 cm, 4 cm, and 6 cm starting from one end.

4. Use the remaining clay to form a cylindrical holder for the wire that is about 3 inches (7.5 cm) tall. You should be able to stick one end of the wire into the side of the clay holder so that when

the wire is held parallel to the table the free end of the wire will be positioned above the candlewick (in the top of the flame when the wick is lit, as shown in Figure 46.1).

5. Place balls of margarine about ½ inch (1.25 cm) in diameter on the marks on the wire.

6. Fold the aluminum foil in half placing the long sides together. Then connect the ends with tape, forming a circle about 4 inches (10 cm) high.

7. Place the circle of aluminum around the candle, wire, and cylindrical clay holder. This will shield drafts from the materials.

8. Use a match to light the candle.

9. With the end of the wire in the tip of the flame, observe the samples of margarine. Make note of the order in which the samples show any sign of melting. When all of the samples show signs of melting, blow out the flame.

Figure 46.1

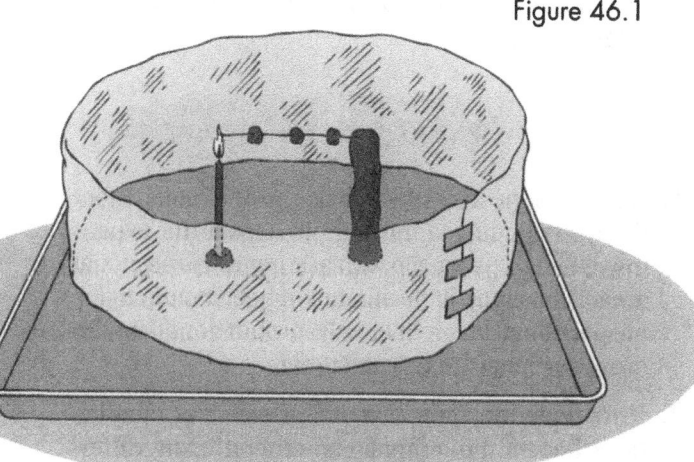

Results

The samples of margarine melt in order of placement from the flame, with the first margarine sample closest to the flame showing the first signs of melting with the sample farthest from the flame melting last.

249

Why?

Thermal conduction is the movement of heat through a substance from a region of high temperature to a region of lower temperature. Conduction occurs in fluids when their moving molecules collide. But conduction is more commonly the method of heat transfer through a solid material without the movement of the solid material itself. Materials through which heat flows readily are called **thermal conductors** or **conductors.** Metals, such as the paper clip in this experiment, are particularly good conductors of heat because of a high concentration of **free electrons** (electrons in some solids, particularly metals, that are attracted relatively equally to all nearby atoms and thus are not tightly bound to a single site and are relatively free to move through the solid), which transfer heat when they collide with the atoms of the metal. The measure of the ability of a material to conduct heat is called **thermal conductivity.**

If a substance is heated, such as the wire in this investigation, the heat is conducted from the heated end to the cooler end. The fact that the margarine sample closest to the heated end of the wire melts first, followed by the second sample, and then the third, shows that it takes time for the heat to be conducted from the hotter end of the wire to the cooler end. Thus the amount of heat conducted through the metal wire from the warmer end to the cooler end is proportional to the time during which conduction has been taking place.

Try New Approaches

1. How does the cross-sectional area of a metal affect the heat conducted through it? Repeat the experiment using a paper clip with a larger cross-sectional area. Determine if the amount of heat conducted is proportional to or inversely proportional to the cross-sectional area of a material.

2. How does the type of metal affect heat conduction? Repeat the original experiment using different types of wire with the same cross-sectional area thus the same **gauge** (a measure of a standard size). Wire strippers have gauge sizes that can be used to measure the circumference of the wire. Generally, a small paper clip is 22 gauge and is made of steel. Use a piece of wire of equal gauge and length but made of different material, such as copper. Be sure to strip away all of the insulation

from the wire. Prepare two clay holders and place them on opposite sides of the candle. Position both wires so they are at the same height and have the same amount of metal in the flame. A large-diameter circle will be needed. Compare the time it takes for each margarine sample to melt. For information about good thermal conductors see Mary Jones's *Physics* (New York: Cambridge University Press, 1997), pp. 82–83.

Design Your Own Experiment

1. Does length of a wire affect heat conduction? Design an experiment to show the effect of length of a conduction material on heat conduction. One way is to repeat step 2 in "Try New Approaches," using two wires of the same gauge but different lengths. When the first sample of margarine shows signs of melting, blow out the candle. Continue to observe the remaining blobs of margarine. Use the results to determine if the same amount of heat is transferred from one end of the wire to the other.

2. How is temperature difference between two materials related to heat conduction between them? Design a way to determine if a larger temperature difference causes more or less heat to flow from one material to another. One way is to fill two Styrofoam cups one-fourth full with water. Put hot tap water in cup A and cold tap water in cup B. Measure the temperatures of the hot and the cold water. Place two metal washers in the hot water. After 3 minutes, use a spoon to remove the washers from the hot water and place them in cup B, the cup of cold water. At the end of 1 minute, stir the water in cup B, then measure the temperature of the water in the cup. Note the difference between the original temperature of the cold water and the temperature after the warm washers had been in the water for 1 minute. Repeat twice, first use warm water in cup B made warm by mixing equal amounts of hot and cold tap water. Then use water in cup B made colder by adding one or more ice cubes to the water. Allow the ice to remain in the water for 2 minutes, stir, then remove the ice before adding the washers. For more information about a material's **temperature gradient** (the temperature change with distance along a mate-

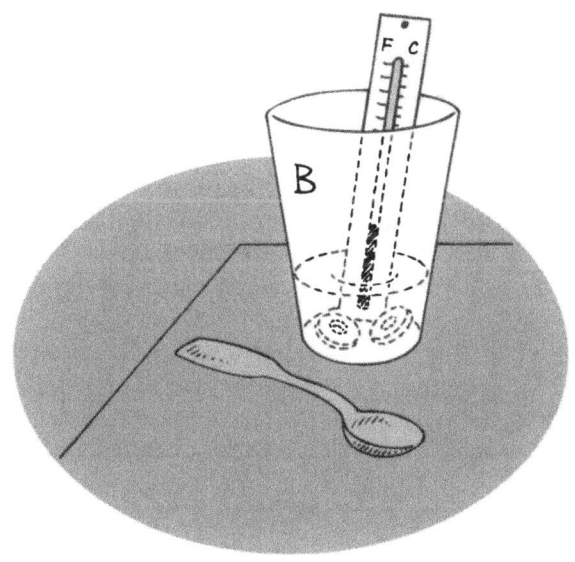

Figure 46.2

rial), see Corine Stockley's *Illustrated Dictionary of Physics,* (London: Usborne, 2000), p. 28.

3. **Insulators** are materials with a low concentration of free electrons, and **thermal insulators** are poor conductors of heat. These materials, such as glass, paper, and Styrofoam, depend on the interaction of vibrating atoms and molecules to transfer heat, which is a less efficient method than by movement of free electrons in conductors. Thus less heat is conducted by thermal insulators. Design an experiment to compare the insulating properties of materials used to hold hot liquids. One way is to use cups each made of different insulating materials. Add equal amounts of hot water to each cup. Place a thermometer in each container. Measure the temperature of the water in each container every 2 minutes for 30 minutes or until no further temperature changes occur. Record the temperature in a Temperature/Insulation Data table like Table 46.1. Use the data to prepare a graph recording the information, using a different color ink for each material on the graph.

Get the Facts

1. *Specific heat* of a substance is the amount of energy that must be added to raise the temperature of a unit mass one temperature unit. When 1 calorie of heat is added to 1 gram of water, the water's temperature rises 1° C. What are the specific heats of other materials, such as aluminum? How can specific heat be used to determine the amount of heat transferred through metal? For information see specific heat in a physics text.

2. There are temperature-sensitive nerve endings in your skin that can detect differences between the temperature inside and outside your body. You perceive an object as feeling cold when heat is transferred from your body to that object. Why do some things that are the same temperature but made from different materials, such as a carpet and a tile floor, feel as if they have different temperatures from each other? For information, see Annabel Craig's *Science Encyclopedia* (London: Usborne, 1988), p. 14.

Materials	TABLE 46.1 TEMPERATURE/INSULATION DATA									
	Temperature, °F (°C)									
	Time, minutes									
	2	4	6	8	10	12	14	16	18	20
Glass										
Paper										
Styrofoam										

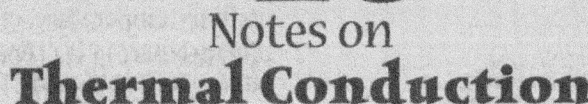

Notes on
Thermal Conduction

Key Facts:

_____

_____

_____

_____

My Results:

_____

_____

_____

_____

_____

_____

_____

_____

Conclusions:

_____

_____

_____

_____

Results of Try New Approaches:

_____

_____

_____

_____

_____

Notes on Designing My Own Experiment:

_____
_____
_____
_____
_____
_____
_____
_____
_____
_____
_____
_____
_____
_____
_____
_____
_____
_____
_____
_____
_____
_____
_____
_____
_____
_____

Notes on Get the Facts:

_____
_____
_____
_____
_____
_____
_____
_____

Electricity is the name given to an effect resulting from the presence of stationary or moving electric charges and the effects they cause. The word *electricity* was coined by William Gilbert (1544–1603), an English physicist and physician known primarily for his original experiments on the nature of electricity and magnetism (the force around a magnet). Rubbing two materials together, such as your feet against a carpet, causes the separation of two kinds of electric charges in the atoms that make up the materials. A buildup of stationary charges are called static charges. If enough charges separate, a spark, called static discharge, is produced when the charges recombine. American scientist and statesman Benjamin Franklin (1706–1790) named the two kinds of charges positive and negative. He also experimentally demonstrated that lightning, like the small spark created when you touch a metal doorknob after rubbing your feet on the carpet, is an example of static discharge, which is a loss of static charges. Static electricity is specifically the effect produced by static charges.

In this project, you will learn about two methods by which materials are charged: friction and conduction. You will discover the effects of electrostatic induction. You will also determine how materials and distance between charged materials affect the electric force.

Getting Started

Purpose: To charge materials electrically by friction and induction.

Materials

9-inch (22.5-cm) round
  balloon
pencil
9-ounce (270-ml) paper cup

tap water
helper
wool cloth (scarf or
  sweater)

Procedure

1. Inflate the balloon to a size that is easily held in one hand. Tie a knot in the neck of the balloon.

2. Use the pencil point to punch a small hole in the side of the cup near its bottom.

3. Ask a helper to hold his or her finger over the hole in the cup, fill the cup with water, then set the cup on the edge of a sink with the hole pointing toward the sink.

4. Rub the balloon on the cloth five or more times.

5. Ask your helper to remove his or her finger from over the hole and observe the direction of the water coming from the hole in the cup.

6. Hold the balloon near but not touching the stream and observe any change in the direction of the stream of water.

Figure 47.1

Results

The stream of water bends toward the balloon.

Why?

Electricity is any effect resulting from the presence of stationary or moving electric charges. A **charge** (**electric charge**) is the property of particles within atoms that causes a **force** (a push or pull on an object) between the particles. The force between particles due to their charges is called an **electric force.** The two forms of charges are called positive and negative. When two like charges (positive and positive or negative and negative) are near each other, they **repel** (push apart) each other. But when two unlike charges (positive and negative) are near each other, they **attract** (pull together) each other. The property of space around a charged object that exerts an electric force on other charged objects is called an **electric field.** The source of positive and negative charges are the protons and electrons in atoms, which contain a nucleus. **Protons** are positively charged particles inside the **nucleus** (central part), and **electrons** are negatively charged particles outside the nucleus.

Physical contact, such as rubbing, between uncharged material is one method, called the **friction method,** of electrically charging them. **Friction** is the name of forces that oppose the motion of two surfaces in contact with each other, such as the rubbing of the balloon and cloth together. Before the balloon and the cloth are rubbed together, they are **neutral** (having an equal number of positive and negative charges, thus having no electric charge). This is because the atoms they are made of have an equal number of protons (positive charges) and electrons (negative charges). Atoms can become charged by either losing or gaining electrons. This happens because electrons, unlike protons, are free to move. If an atom loses an electron, the atom then has more positive charges (protons) than negative charges (electrons), and is therefore positively charged. If an atom gains an electron, it has more negative charges (electrons) than positive charges (protons), and is consequently negatively charged.

When two objects are rubbed together, one of them tends to lose electrons more than the other. The loss of electrons in one object results in its becoming positively charged, and the gain of electrons by the other object results in that object becoming negatively charged. In this experiment, when the rubbing stops and the balloon and cloth are separated, the electric charges stop moving. The balloon has a negative charge and the cloth a positive charge. An object with more of one kind of charge than another is said to be **charged.** These electric charges remain stationary, thus are called **static charges** (a buildup of stationary electric charges). **Static electricity** is the effect of static charges.

While water molecules are neutral, they are **polarized,** meaning their positive and negative charges are separated so that they have a positive and negative end. The presence of the negatively charged balloon causes the negative end of the water molecules to be repelled. Because water is a liquid, its molecules have more freedom of motion, so the water molecules rotate until the positive ends of the molecules are facing the balloon. Thus the side of the stream of water facing the balloon becomes more positively charged. Since unlike charges are attracted to one another, the positively charged water stream is attracted to the negatively charged balloon. The process of polarizing a neutral material by separating its positive and negative charges due to the proximity (nearness) of a charged object is called **electrostatic induction.** The rotation of the polarized water molecules by the charged balloon is an example of electrostatic induction.

Try New Approaches

How would using a positively charged object affect the results? Since the balloon stripped away electrons from the wool scarf, the scarf becomes positively charged. Repeat the original experiment, again rubbing the balloon with the scarf, but this time using the scarf instead of the balloon.

Design Your Own Experiment

1. Two charged objects that have different charges attract each other, while two charged objects that have the same charge repel each other. So if a negatively charged object is brought near a neutral, nonpolarized solid, will the solid be polarized by induction because of the separation of its positive and negative charges? Design a way to determine if neutral solids can be polarized by electrostatic induction. One way is to prepare a pendulum using a bob (the suspended object on a pendulum) made of aluminum foil. Crush a small piece of aluminum foil to make the bob, then tie it to a 12-inch (30-cm) piece of string. Using tape, secure the free end of the string to the edge of a table. Charge a balloon as in the original experiment by rubbing it on a wool scarf. Hold the charged balloon under the hanging aluminum bob so that the charged balloon is near but not touching it. Slowly move the charged balloon to

the left, then move it to the right. Observe the motion of the hanging bob.

Figure 47.2

2a. Electrical conduction is the movement of electric charges through a substance. **Electrical conductivity** is the measure of the ability of a substance to conduct an **electric current** (the flow of electric charges). Conductors or **electrical conductors,** such as metals, are substances with a high concentration of free electrons and are good conductors of electric charges. **Insulators** or **electrical insulators** are substances with a low concentration of free electrons and are poor conductors of electric charges. **Charging by conduction** is the method of charging a neutral object by touching it with a charged object. Design an experiment to determine which would hold a static charge longer, a conductor such as aluminum, or an insulator such as paper. One way is to prepare two sets of pendulums: set one with two aluminum bobs, and set two with two paper bobs equal in size to the aluminum bobs. Use the procedure in the previous experiment to make the four bobs. First hang the pendulums from set one by taping the free ends of their strings to a table so that the bobs are slightly separated. Charge the aluminum bobs by conduction. Do this by touching them simultaneously for 1 to 2 seconds with a charged balloon. The bobs will

have like charges, so they will separate. Measure the length of time of separation between the bobs. Make note of how many times you rub the balloon against the cloth when charging it. Then repeat the procedure with the paper bobs from set two.

b. Electric discharge is the loss of static electricity. Does the magnitude of the charge on the balloon affect electric discharge? Repeat the experiment twice, first rubbing the balloon more times, then rubbing it fewer times against the cloth. Make sure that all materials start with a neutral charge. Allowing the bobs to hang undisturbed and away from charged materials for 10 or more minutes should give them time to electrically discharge. This happens as excess electrons on the bobs are picked up by molecules in the air, especially water molecules.

c. What effect would bobs made of insulating material have on the results? Repeat the experiment using insulating materials, such as paper and cotton. Be sure that the only variable you change is the type of material used; keep the size and shape of the bobs as similar as possible.

3. An instrument called an **electroscope** is used to determine the presence of an electric charge on an object. As shown in Figure 47.3, an electroscope has a conducting metal rod with two thin metal leaves attached. If the rod is charged by conduction, the leaves separate. Design a way to determine any difference in the effect of charging the electroscope with a positively or negatively charged object. One way is to build an electroscope and test it first with a positively charged object (wool scarf after it has been rubbed with a balloon) then with a negatively charged object (balloon after it has been rubbed on a wool scarf). The electroscope can be made using a glass jar with a circle of cardboard to cover its top. Use pliers to reshape a large paper clip into a U-shaped loop. Punch two holes in the center of the cardboard circle about ½ inch (1.25 cm) apart. Push the ends of the loop of wire through the holes and mold a small piece of modeling clay around the base of the loop to hold the wire in place. The bent section of the loop should stand up above the cardboard about ½ inch (1.25 cm). Bend both ends of the wire loop outward to form hooks. Hang a small light-weight aluminum foil strip on each wire hook. The strips should jiggle back and forth freely. If they do not, enlarge their holes. Place the

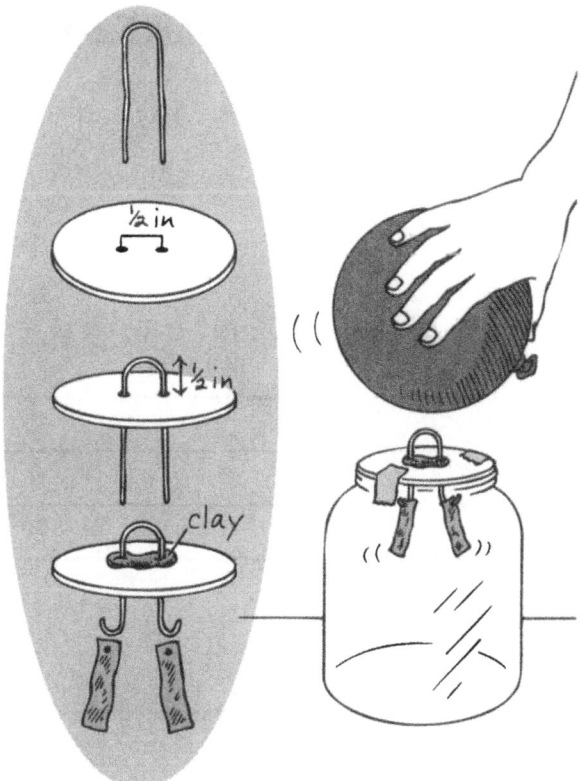

Figure 47.3

cardboard circle over the mouth of the jar with the foil strips hanging down inside the jar. Tape the cardboard cover to the jar. Charge the foil strips by conduction (touching the metal top with a charged object). Variables that could affect the results of your experiment include size and weight of the aluminum foil strips, humidity, and separation of the foil strips. Electroscopes can also be charged by induction. See a physics book for information on this method of charging as well as instructions on how to charge an electroscope by induction.

Get the Facts

1. Static electricity has many uses, for example, electrostatic copying machines make duplicates of printed material by attracting negatively charged particles of powdered ink to positively charged paper. But static electricity also causes problems, such as sparks produced in a grain bin that can cause explosions. Find out more about the positive uses and negative problems of static electricity. For information see Mary and Geoff Jones's *Physics* (New York: Cambridge University Press, 1997), pp. 63–68.

2. In 1785, French physicist Charles Coulomb (1736–1806) used a type of balance to measure the force between two charged spheres. Use a physics text to find out more about Coulomb's experiment and the equation he used to describe the relationships among electric forces, charge, and distance. This equation is called *Coulomb's law.* Another source of information is Karl F. Kuhn's *Basic Physics: A Self-Teaching Guide.* (New York: Wiley, 1996), p. 139.

3. The triboelectric series ranks materials according to the amount of energy needed to remove their electrons. This listing provides information about predicting the charges of two materials that are rubbed together. For information about the triboelectric series see Kuhn's *Basic Physics*, p. 143.

Notes on
Static Electricity

Key Facts:

_____

_____

_____

_____

My Results:

_____

_____

_____

_____

_____

_____

_____

Conclusions:

_____

_____

_____

_____

_____

Results of Try New Approaches:

_____

_____

_____

_____

_____

Notes on Designing My Own Experiment:

Notes on Get the Facts:

48 Series Circuit
Sequential Path

An electrical circuit is the path that electric charges follow. When there is only one path through which an electric current can flow, the circuit is called a series circuit. If any part of a series circuit is broken, then no current can flow through any part of the circuit. Some holiday lights are in series circuits, so when one light burns out the circuit is broken and all the lights stop working.

In this project, you will use a model to demonstrate a series circuit. You will determine how to measure the voltage, current, and resistance of a series circuit. You will also learn how to determine resistance mathematically, using Ohm's law.

Getting Started

Purpose: To demonstrate a series circuit.

Materials

1.5-volt battery
battery holder with insulated wires (red and black)

flashlight lamp (for E-10 screw-base holder)
lamp holder with E-10 screw-base

Procedure

1. Place the battery in the battery holder so the battery's negative terminal is at the end with the holder's black wire.

2. Screw the lamp into the lamp holder.

3. Holding the insulated part of the wires attached to the battery holder, touch the bare ends of the wires to the screws on either side of the lamp holder (see Figure 48.1). **Caution:** *Do not leave the wires on the screws for more than a few seconds. The bare wire and lamp can get hot enough to burn you. Allow them to cool before touching them.*

4. Observe the lamp when only one wire leading from the negative terminal of the battery touches a screw on the lamp holder.

5. Repeat step 4 using only the wire from the positive terminal.

Figure 48.1

Results

The lamp glows only when the two wires leading from the battery touch the two screws on the lamp holder, one wire on either side of the lamp holder.

Why?

An electric current is the flow of electric charges. Electric current moves through conductors, such as the connecting wires from the battery holder and the metal parts of the lamp holder, lamp, and **battery** (a device that uses chemicals to produce an electric current). A path made of conducting materials through which an electric current travels is called an electric circuit. If the electric circuit is a loop, meaning a continuous path, it is called a closed circuit. If there is a separation of the conducting material forming the electric circuit, it is called an open circuit. If the circuit, like the one in this experiment, has only one path through which an electric current can flow, it is a **series circuit.**

The lamp indicated whether the circuit was open or closed. For the lamp to glow, electrons must move through its **filament,** which is a thin coil of wire inside the lamp. Because the filament wire is small, the electrons flowing through it are more likely to collide with atoms in the wire. These collisions cause the atoms to vibrate, thus increasing the temperature of the wire. When there is enough current, the wire heats enough to glow. When the circuit was closed, the wires attached to either side of the lamp led to each end of the battery allowing electrons to flow through the circuit, and the lamp glowed. An open circuit was formed when one of the wires from the lamp was removed from the battery. It made a break in the circuit, so the electrons could not flow. The light did not glow when the circuit was open.

The battery provides **direct current (DC)** (electric current moving in one direction). The arrows indicate that in a closed circuit, the current, indicated by the symbol e′ in Figure 48.2, moves away from the negative terminal of the battery, through the lamp, and back to the positive terminal of the battery.

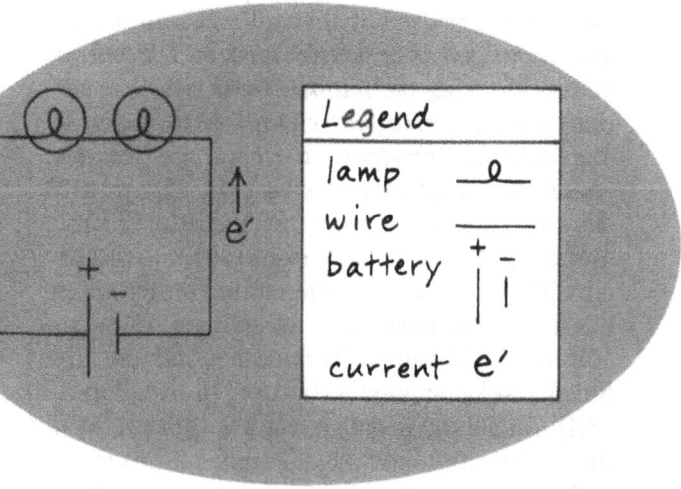

Figure 48.2

Try New Approaches

1. Would it affect the lamp's brightness if the electricity flowed in a reverse direction through the lamp? Repeat the experiment, rotating the battery holder 180° so that the positive (red wire) and negative (black wire) terminals are reversed.

2a. Would adding more lamps in the series circuit affect the brightness of each lamp? Repeat the original experiment, using two lamps in sequence. Use a 4-inch (10-cm) piece of 22-gauge wire to connect the lamp holders. Use wire cutters to strip a small section from each end of the wire and attach the bare wires to one screw on each of the lamp holders. **Science Fair Hint:** Use a schematic drawing, such as the one in Figure 48.2, to show a two-lamp series circuit.

b. When lamps are connected in a series, what happens when one burns out? Determine this by unscrewing one of the lamps that is connected in series. (*Note:* When a lamp burns out, the filament in the lamp breaks. Thus unscrewing the lamp gives the same effect as using a burned-out lamp.)

3. What effect does an increase in the number of batteries used have on the brightness of the bulbs? Repeat the original experiment, using one bulb and connecting two 1.5-V batteries in series by using a wire to join the negative terminal of one battery to the positive terminal of the other. **CAUTION:** *Use only 1.5-V batteries. Connecting more than two 1.5-volt batteries in a series or using batteries with a greater voltage may burn out the lamp as well as produce a dangerous amount of current.*

Design Your Own Experiment

1a. **Coulomb (C)** (charge on 6.25×18^{18} electrons) is the SI unit for quantity of electric charge. One coulomb per second is called an **ampere (A)** (unit measure of electric current), more commonly called amps. A device used to measure the amount of electric current in a circuit is called an **ammeter.** **A multitester** is an instrument that has the ability to work like a number of instruments, including an ammeter and a **voltmeter** (an instrument used to measure voltage). A multitester can be purchased at an electronics store. Design an experiment that uses a multitester as an ammeter to measure the current in any or all of the series circuits in the previous experiments. Follow the directions provided with your multitester, but note the information shown here for the multitester used by the author. **Caution:** *An ammeter is always connected in series. This means that when using the multitester to measure current, you must break the circuit being measured and make the ammeter part of the circuit. You can ruin the multitester if this is not done.* Figure 48.3 shows a diagram of a circuit with a battery, lamp, switch (device used to open and close an electric circuit), and ammeter. The author's multitester has a scale for measuring DC current from 0 to 150 mA (milliamps). To measure current, the function selector is set to 150 mA DC (see Figure 48.4). Attach the negative test lead (black) from the multitester to the negative side of the circuit and the positive test lead (red) to the positive side of the circuit, as shown in Figure 48.4. Only the DCV/mA part of the scale on a multitester is shown in Figure 48.4. While numbers are not printed on this scale, with the tester selector set to 150 mA DC, the fifteen long marks on the scale each measure 10 mA, which is read as 10 milli-amps. The current reading for the circuit in Figure 48.4 shows the scale needle on the small mark between 30 mA and 40 mA; thus the current is 35 mA and is equal to 0.035A.

b. Does it matter where in the circuit the multitester is placed? Design series circuits and use the multitester to test in different places along the circuit, such as between two lamps or on the positive side of the battery and then on the negative side.

2. The driving force that pushes the electrons around the circuits in the experiments in this chapter is the battery, which acts like an electron pump. Voltage is a measure of the amount of potential energy that the battery transfers to electrons in a circuit. It is a measure of the difference in the energy on either side of the cell.

The unit measure of voltage is volts (V). A multitester can act as a voltmeter. Design an experiment so that voltage can be measured. You may wish to place a switch in the circuit to make it easier to open and close the circuit. Use the multitester to measure the voltage in any or all of the series circuits in the previous experiments. Follow the directions provided with your multitester, but note the information shown here for the multitester used by the author. **CAUTION:** *A voltmeter should NEVER be part of a circuit; instead, it should be connected across a circuit.* Set the function selector to the lowest DC V position, which is 15 on the author's multitester. To determine the voltage through one of the lamps, touch the negative test lead (black) to the negative side of the circuit and the positive test lead (red) to the positive side of the circuit, as shown in Figure 48.5A. The voltage in the diagram is read as 1.5 volts. Figure 48.5B shows a schematic for the circuit with a voltmeter.

Figure 48.3

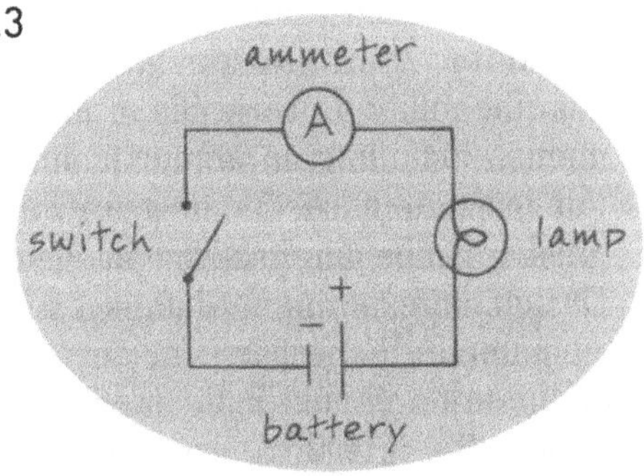

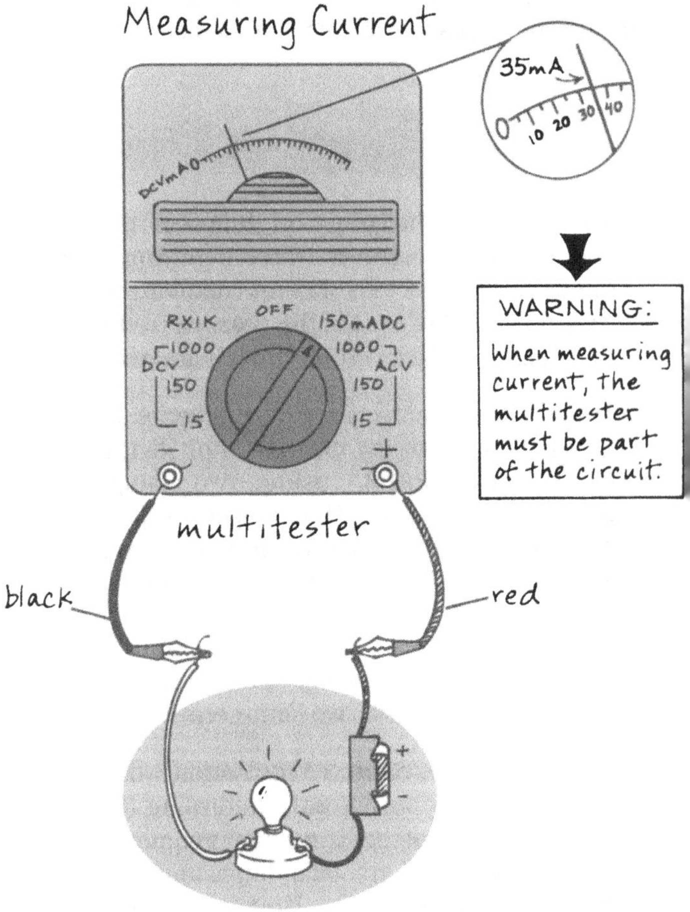

Figure 48.4

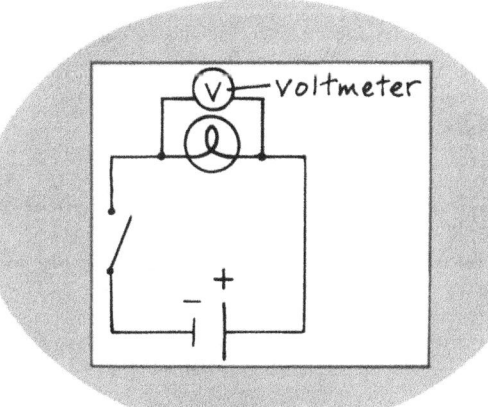

1.5 V

WARNING:
When measuring voltage, the multitester is NEVER in series with the circuit.

black   red

Figure 48.5A

Figure 48.5B

voltmeter

3. Each device in a circuit affects the flow of electrons, and some restrict the flow more than others. Any device in a circuit, such as a lamp and wire, offers resistance. But a device that is used to create electrical resistance in an electric circuit is called a **resistor.** Resistance is measured in the SI unit of **ohm (Ω).** A multitester can be used to measure resistance, but the resistance of the circuits in this chapter may be too low to be measured accurately by the tester. Instead, since the voltage and the current can be measured, the relationships among voltage, current, and resistance, known as **Ohm's law,** can be used to calculate resistance. This relationship is expressed as $V = I \times R$, which is read: V voltage (in volts) equals I, current (in amperes) times R, resist-

ance (in ohms). If the voltage and the current are known, the resistance can be calculated using the following formula:

$$R = V \div I$$

For example, if a circuit has a voltage of 1.5 V and a 35 mA current, the resistance would be:

$$R = 1.5\,V \div 0.035\,A$$
$$= 42.85\,\Omega$$

For more information about Ohm's law, see Karl F. Kuhn, *Basic Physics: A Self-Teaching Guide* (New York: Wiley, 1996), pp. 152–153.

Get the Facts

A battery is made of one or more *electrical cells* connected in series. What are electrical cells made of? Is a flashlight battery technically a battery or a cell? How do electrical cells produce the driving force of the battery? What is a dry cell? What is the electric potential of a battery, and what causes it? For information, see physics texts as well as Mary and Geoff Jones, *Physics* (New York: Cambridge University Press, 1977), pp. 202–203.

48
Notes on
Series Circuit

Key Facts:

_____
_____
_____
_____

My Results:

_____
_____
_____
_____
_____
_____
_____

Conclusions:

_____
_____
_____
_____

Results of Try New Approaches:

_____
_____
_____
_____

Notes on Designing My Own Experiment:

_____
_____
_____
_____
_____
_____
_____
_____
_____
_____
_____
_____
_____
_____
_____
_____
_____
_____
_____
_____
_____
_____
_____

Notes on Get the Facts:

_____
_____
_____
_____
_____
_____

Parallel Circuit
Divided Pathways

A parallel circuit is an electric circuit in which the electric current has more than one path to follow. The advantage is that, like adding another lane on a busy freeway, more traffic can flow. With a parallel circuit, more electric current can flow.

In this project, you will determine the path that electrons follow in a parallel circuit. You will measure the total current and voltage of each circuit as well as the connected branches, and you will use these measurements to confirm Ohm's law. You will also investigate how battery cells are connected in parallel and the effect of parallel cells on the current and voltage of a circuit.

Getting Started

Purpose: To determine the path of the electric current in a parallel circuit.

Materials

1.5-volt battery
battery holder with insulated wires (red and black)
2 identical flashlight lamps (for E-10 screw-base holder)
2 lamp holders with E-10 screw-base

wire cutter
12-inch (30-cm) piece of 22-gauge insulated wire
screwdriver with head type for lamp holder screws

Procedure

1. Place the battery in the battery holder.

2. Screw the lamps into the lamp holders.

3. Use the wire cutter to cut the wire into two 6-inch (15-cm) parts. Then strip about ½ inch (1.25 cm) of insulation from each end of the wires.

4. Attach the wires to connect the lamp holders, as shown in Figure 49.1.

5. Holding the insulated part of the wires attached to the battery holder, touch the bare ends of the wires to the screws on either side of one of the lamp holders, as shown in Figure 49.1. Compare

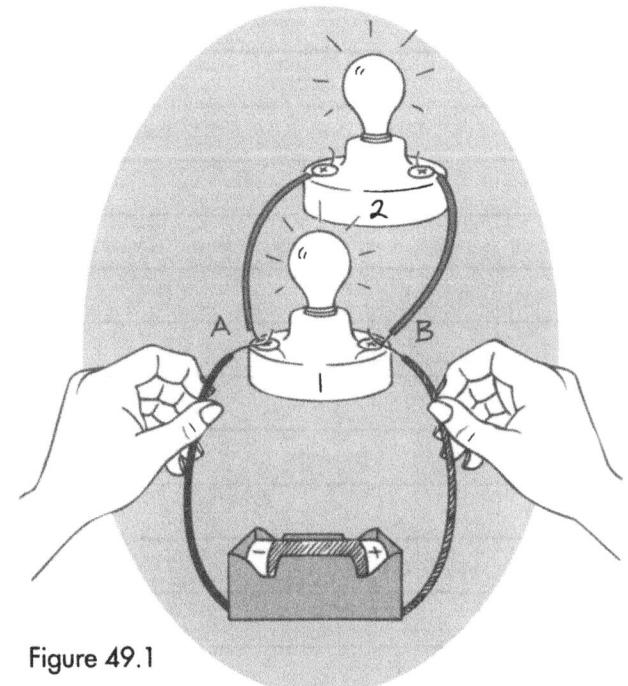

Figure 49.1

the brightness of the lamps. **CAUTION:** *Do not leave the wires on the screws for more than 5 to 6 seconds. The bare wire and lamp can get hot enough to burn you. Allow them to cool before touching them.*

Results

The lamps glow with equal brightness.

Why?

The electric current has more than one path to follow through the connected lamps in this experiment, so the circuit formed what is called a **parallel circuit.** The arrows in the electrical schematic in Figure 49.2 indicate the movement of the current (e′) away from the negative terminal of the battery through the bulbs, then back to the positive terminal of the battery. At junction A (one of the screws on the side of lamp base 1), the current divides before moving through the lamps. Then the current recombines at junction B (one of the screws on the opposite side of lamp base 1) and returns to the positive terminal of

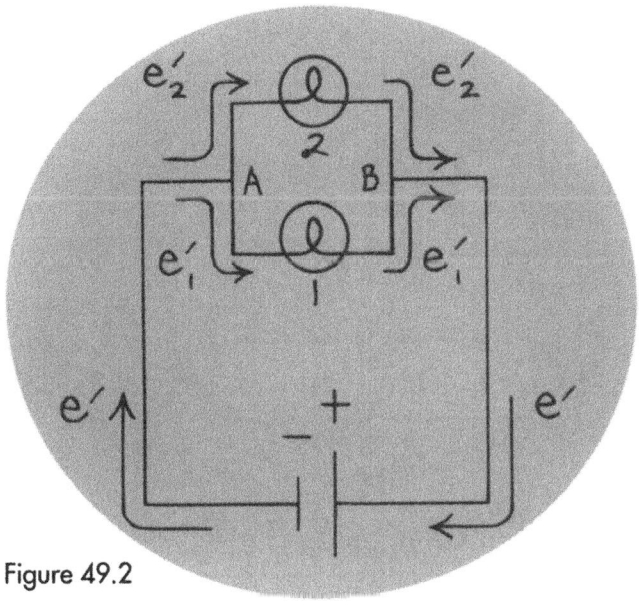

Figure 49.2

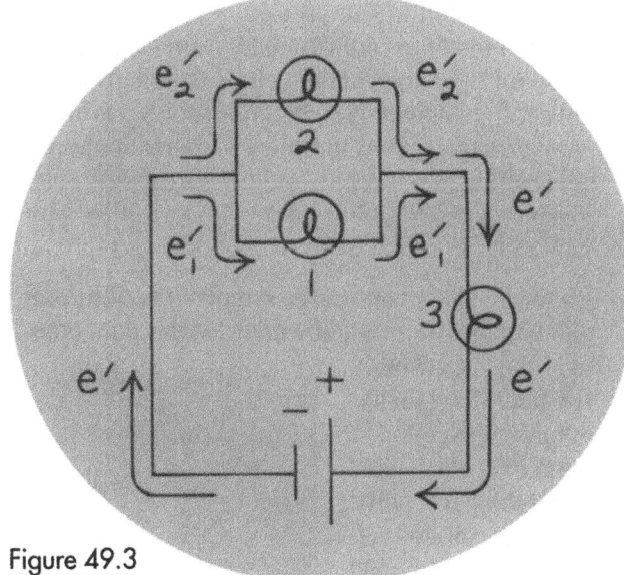

Figure 49.3

the battery. The facts that the lamps are identical and glow with equal brightness show that the same amount of current reaches both lamps.

Try New Approaches

1. Would it affect the brightness of the lamps if more lamps were connected in parallel? Repeat the experiment, adding more lamps.

2. Would it affect the brightness of the lamps connected in parallel if a lamp is placed in series with them? Repeat the original experiment, connecting two lamps as in the original investigation and adding a third lamp in series, as shown in the schematic in Figure 49.3.

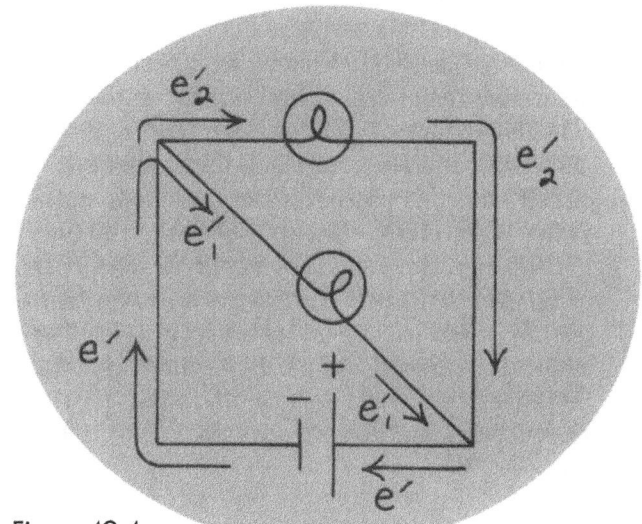

Figure 49.4

Design Your Own Experiment

1. Do the lamps in a parallel circuit have to be geometrically parallel to each other, or just connected so that the electrons have different paths from the negative side to the positive side of the battery? Design different electrical schematics, such as the one shown in Figure 49.4 Then assemble the circuits using the indicated lamps and battery for each schematic. Allow the brightness of the bulbs to indicate any change in the flow of the current through them.

2a. Ohm's law describes the total current in a parallel circuit (I_t) as the sum of the currents in each branch of the circuit. Design a circuit with two or more lamps in parallel to confirm that for devices in parallel $I_t = I_1 + I_2 + I_N$. In the equation, It is the total current, I_1 is the current through lamp 1, I_2 is the current through lamp 2, and I_N represents the sum of other lamps in parallel, such as I_3, I_4, and so on. Use a multitester, such as the one described in Chapter 48, to measure the currents.

b. Ohm's law describes the voltage of lamps in parallel as the same as the total voltage of the circuit.

Use the circuit in the previous experiment and the multitester to confirm that V_t (total voltage) + V_N (sum of voltage across other lamps in parallel) = V_1 (voltage across lamp 1) + V_2 (voltage across lamp 2). For more information about parallel circuits, see Karl F. Kuhn, *Basic Physics: Self-Teaching Guide* (New York: Wiley, 1996), pp. 152–158.

3. Battery cells in series are connected so that the anode (positive terminal) of one cell is connected to the **cathode** (negative terminal) of another cell. In parallel cells, like terminals are connected, anode to anode and cathode to cathode. Figure 18.5 shows two bat- 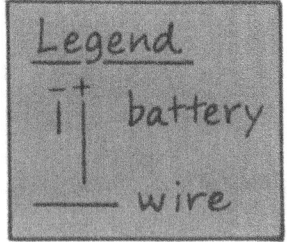 teries connected in series and two batteries connected in parallel. Design an experiment to determine the effect of parallel cells on the current and voltage of a circuit. One way is to connect two batteries in parallel, then assemble a circuit with this parallel battery, a lamp, and a switch. **CAUTION:** *Use only 1.5-V batteries. Connecting more than two 1.5-volt batteries in a series or using batteries with a greater voltage may burn out the lamp as well as produce a dangerous amount of current.* For more information about batteries, see Mary and Geoff Jones, *Physics* (New York: Cambridge University Press, 1997), pp. 202–203.

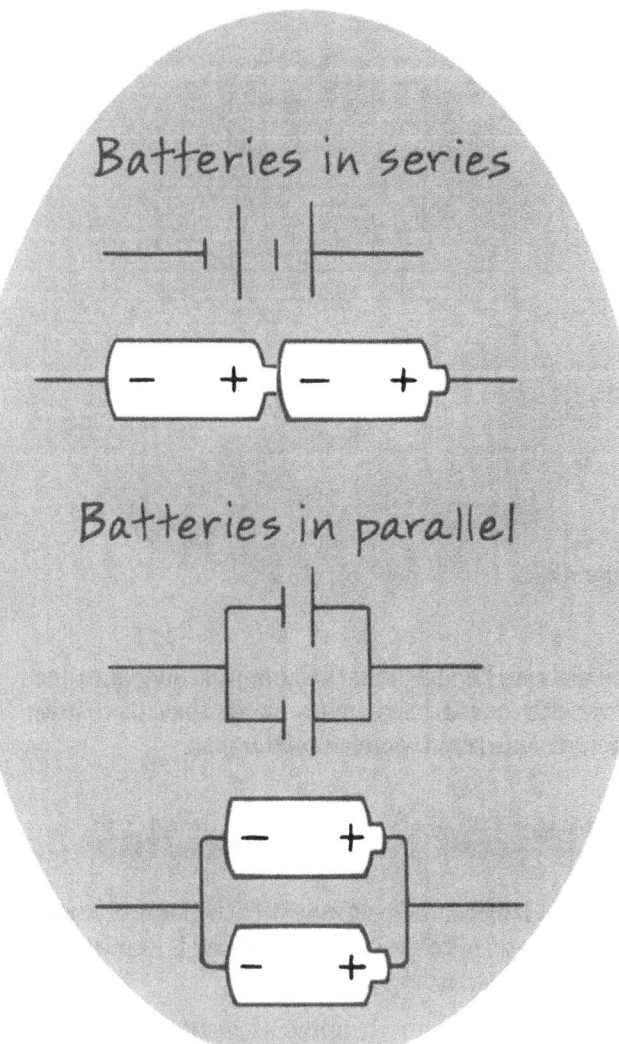

Figure 49.5

Get the Facts

Christmas lights were at one time wired in series to save on cost, but most are now wired in parallel. What effect did a burned-out bulb have when the bulbs were in series? In parallel? Which is used in homes—series or parallel circuits? For information, see P. Erik Gundersen, *The Handy Physics Answer Book* (Detroit: Visible Ink, 1999), pp. 313–314.

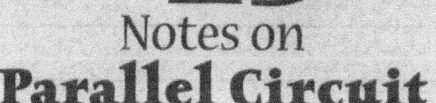

Notes on
Parallel Circuit

Key Facts:

_____
_____
_____
_____

My Results:

_____
_____
_____
_____
_____
_____
_____

Conclusions:

_____
_____
_____
_____
_____

Results of Try New Approaches:

_____
_____
_____
_____
_____

Notes on Designing My Own Experiment:

_____
_____
_____
_____
_____
_____
_____
_____
_____
_____
_____
_____
_____
_____
_____
_____
_____
_____
_____
_____
_____
_____
_____
_____

Notes on Get the Facts:

_____
_____
_____
_____
_____
_____
_____

50 Electromagnetism
Magnetism from Electricity

In 1820, the Danish physicist Hans Oersted (1777–1851) noticed that a compass needle was deflected at right angles to a wire carrying an electric current. Since a compass needle is deflected by a magnet, he concluded that the electric current somehow caused a magnetic field around the wire. Further studies by Oersted proved that any wire carrying an electric current has a magnetic field around it. Oersted's discovery started the study of electromagnetism and the use of an electromagnet (device that uses an electric current to produce a concentrated magnetic field).

In this project, you will determine the direction of a magnetic field around a current-carrying wire. You will determine the effect of the direction of an electric current in a current-carrying wire on the pattern of the magnetic field around it. You will also make an electromagnet, determine its polarity, and test its strength.

Getting Started

Purpose: To determine the direction of a magnetic field around a current-carrying wire.

Materials

pencil
ruler
sheet of copy paper
compass
1.5-volt D battery

wire stripper
18-inch (45-cm) piece of 20- or 22-gauge insulated single-strand wire

Procedure

1. Use the pencil and the ruler to draw two perpendicular lines in the shape of a plus sign across the center of the paper. Label the longer line N, S, and the short one E, W as shown in Figure 50.1.

2. Place the paper on a wooden table and set the compass in the center of the paper, where the lines cross.

3. Allow the needle to come to rest in line with Earth's magnetic field.

4. First rotate the compass so that N on the compass is in line with the needle pointing north. Then lift the compass and rotate the paper so that its compass directions match those on the compass. Replace the compass on the paper.

5. Place the battery on the east side of the paper, with the negative terminal of the battery pointing south.

6. Using the wire stripper, remove about 1 inch (2.5 cm) of insulation from each end of the wire.

7. Bend the wire so that it has a straight center piece slightly wider than the battery's length.

8. Holding the insulated part of the end of the wire, position the wire so that the center section is about 1 inch (2.54 cm) above the compass, and across the compass from north to south. Then touch the bare ends of the wire to the ends of the battery for about 1 second. Notice the direction in which the north end of the compass needle moves. **CAUTION:** *For your safety, hold the insulated part of the current-carrying wire so that you do not get burned or shocked. Holding the wires against the battery terminals for longer than 3 seconds can result in the wire becoming hot enough to burn your skin.*

Figure 50.1

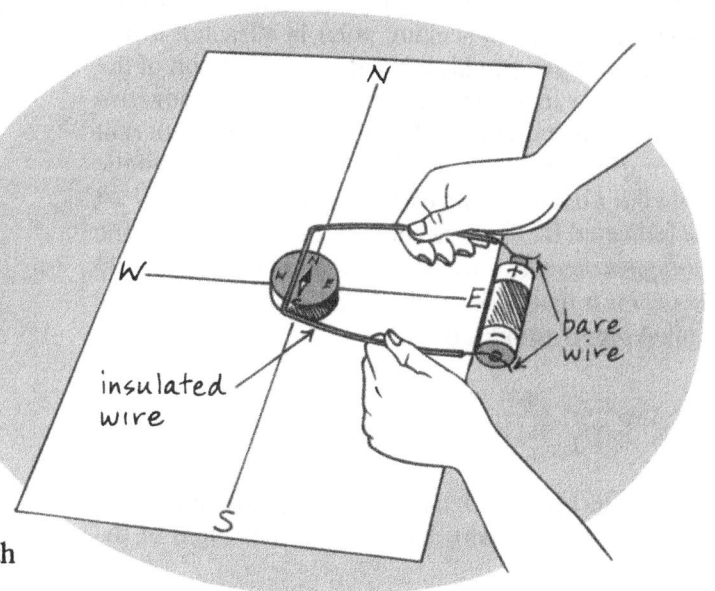

271

9. Move the battery to the opposite (west) side of the compass. With the negative terminal of the battery pointing south as before, repeat step 8.

Results

With the battery's negative terminal pointing south, and the battery on the east side of the compass, the north end of the compass needle deflects from its north position toward the east.

Why?

An electric current is the flow of electric charges through a conductor such as a metal wire. The motion of electric charges also produces a force called a **magnetic force.** The property of a space in which a magnetic force can be detected is called a **magnetic field.** A magnetic field is made up of imaginary lines called **magnetic field lines** that indicate the direction and magnitude of the field. The straight current-carrying wire in this experiment produces a magnetic field as indicated by the movement of the compass needle. The direction of the magnetic field produced by the the current-carrying wire is the direction to which the north pole of the compass points when placed in the magnetic field. The compass needle is a **magnet** (a material with magnetic forces that attracts other magnetic material and is attracted to another magnet). Every magnet has two **magnetic poles** (the regions around a magnet where the magnetic forces appear strongest) called south and north poles. For a freely swinging straight magnet, such as the compass needle, its **south pole** (south-seeking pole) is attracted in the direction of geographical South. Its **north pole** (north-seeking pole) is attracted in the direction of geographical North. The direction of the current flow in the wire is from the battery's negative terminal to its positive terminal. With the current flowing in a south-to-north direction, the magnetic field lines below the wire are directed toward the east as indicated by the deflection of the north end of the compass needle toward the east. The relationship between a magnetic field and an electric current is called **electromagnetism.**

Try New Approaches

What effect would changing the direction of the current through the wire have on the deflection of the

compass needle? Rotate the battery 180° so that the terminals of the battery have been reversed.

Design Your Own Experiment

1. A straight current-carrying wire is said to have magnetic field lines encircling it. Design a way to show that the direction of the magnetic field lines are in a circle around a current-carrying straight wire. One way is to compare the direction of the magnetic field above and below the wire in the original experiment. Design a way to raise the compass and place the wire below it, such as forming a stand for the compass by bending the ends of an index card to form a table shape.

2. Design a way to show the pattern of the magnetic field lines around a magnet. One way is to place a piece of insulated wire through a piece of cardboard, such as the top of a small cardboard box. Sprinkle a thin layer of iron filings on the cardboard around the wire in a circle with about a 4-inch (10-cm) diameter. Connect the ends of the wire to the terminals of a 1.5-volt battery. Observe the pattern of concentric circles (circles with a common center) around the wire.

3. An **electromagnet** is a device that uses electric current to produce a concentrated magnetic field. An electromagnet is made of a **solenoid** (coil of wire through which a current can pass) with a core of magnetic material such as iron. The current-carrying wire in a solenoid produces a magnetic field, which **magnetizes** (causes a substance to become a magnet) the iron core. Design an experiment to determine the **polarity** (the direction of the magnetic poles) of an electromagnet. One way is to wrap a 3-foot (90-cm) piece of 22-gauge insulated wire around a 16d finishing nail (also called a 16-penny nail) (see Figure 50.2). Leave about 4 inches (10 cm) of free wire at each end. Use a wire cutter to strip about 1 inch (2.5 cm) of insulation from the ends of the wire. Allow the compass to align with Earth's magnetic north. With a 1.5V battery in a battery holder, twist together the bare end of one solenoid wire and the bare end of one battery-holder wire. Hold the electromagnet so that the pointed end of the nail is near but not touching the west side of the compass. While in this position, touch the free solenoid wire and the free battery wire together for 1 second. Note the direction in which the north end of the compass needle moves. If the end of the nail pointing toward the compass attracts the north end of the compass needle, the end is the south pole of the electromagnet. If the north end of the needle is repelled, the nail's end is the north pole of the electromagnetic. Reverse the direction of the battery and repeat the procedure.

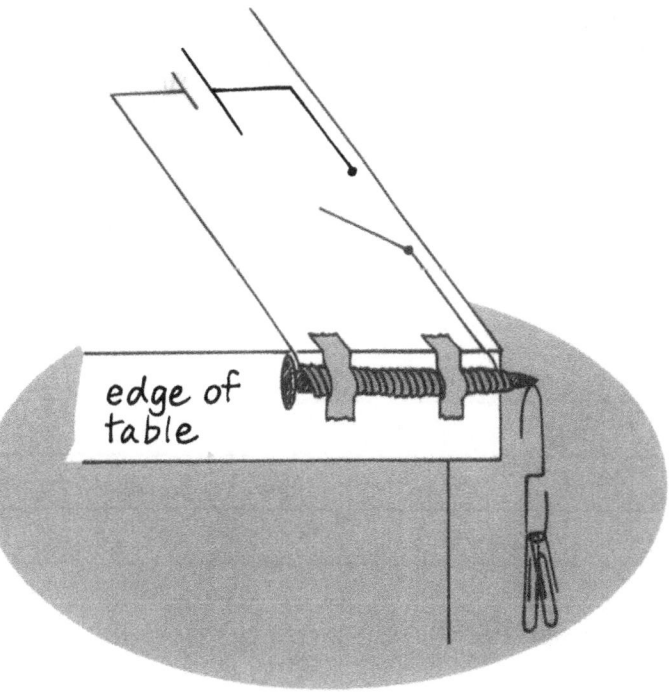

Figure 50.3

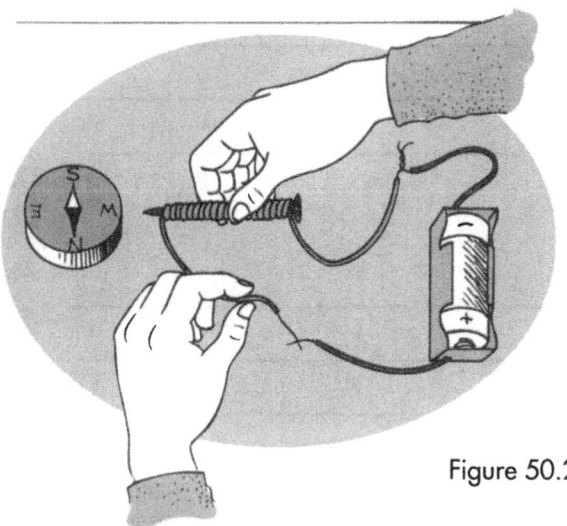

Figure 50.2

4. How does the number of wire coils in an electromagnet affect the strength of its magnetic field? Design an experiment to test the magnetic strength of an electromagnet. One way is to use the electromagnet from experiment 3 made of 3 feet (90 cm) of insulated wire. Assemble a circuit using the electromagnet, a 1.5-volt D battery in a battery holder, and a switch. Tape the electromagnet to the edge of a wooden table as shown in Figure 50.3. Use metal paper clips to test the strength of the electromagnet. Bend one paper clip to form a hook that other paper clips can be hung on. Close the switch and touch the paper clip hook to the pointed end of the nail. Add paper clips to the hook one at a time until the weight of the clips causes the hook to pull away from the nail. Then repeat the experiment using twice as much—6 feet (180 cm) of wire to make the electromagnet. If all the coils will not fit on the nail, wind them as tightly as possible, then wind the next layer over the top, still turning in the same direction. **CAUTION:** *If you feel any*

warmth through the insulated area of the nail, open the switch. Do not touch the bare nail or bare ends of the wire because electric current flowing through the wire can cause these areas to get hot enough to burn your skin.

Get the Facts

1. Television images are the result of thousands of electrons hitting the television screen. What effect do electromagnets play in the direction in which the electrons move? For information see P. Erik Gundersen's *The Handy Physics Answer Book* (Detroit: Visible Ink, 1999), pp. 329–330.

2. *MAGLEV* stands for "magnetically levitated." How are MAGLEV trains different from conventional trains? For information see Gundersen's *The Handy Physics Answer Book*, pp. 330–331.

3. *Left-hand rules* are used to find the force on current or moving particles in a magnetic field. They are also used to find the direction of a magnetic field caused by current in straight wires as well as in solenoids. What are the left-hand rules? How do left-hand rules compare? For information see physics texts.

NAME

Notes on
Electromagnetism

Key Facts:

_____
_____
_____
_____

My Results:

_____
_____
_____
_____
_____
_____
_____

Conclusions:

_____
_____
_____
_____
_____

Results of Try New Approaches:

_____
_____
_____
_____
_____

Notes on Designing My Own Experiment:

_____
_____
_____
_____
_____
_____
_____
_____
_____
_____
_____
_____
_____
_____
_____
_____
_____
_____
_____
_____
_____
_____
_____
_____

Notes on Get the Facts:

_____
_____
_____
_____
_____
_____

51 Static Fluids
Fluids at Rest

Static means no motion, and fluids are liquid or gaseous materials that can flow. Thus static fluids is the study of the characteristics of liquids and gaseous materials that are at rest.

In this project you will experimentally compare air pressure in different directions at a given point. You will experimentally determine the relationship between pressure and depth in a static liquid. You will also experimentally test Pascal's law, which states that pressure applied to an enclosed fluid is transmitted equally in all directions and to all parts of the enclosing vessel.

Getting Started

Purpose: To compare air pressure in different directions at a given point.

Materials

1-pint (500-ml) jar
tap water
1 index card
large bowl

Procedure

1. Fill the jar with water.
2. Cover the mouth of the jar with the card.
3. Hold the jar over the bowl.
4. With one hand over the card, invert the jar.
5. Carefully remove your hand from the card.
6. Keeping the mouth of the jar at relatively the same height, slowly rotate the jar through a complete 360° circle, so that the mouth of the jar faces down, then up, and then down again.
7. Observe the surface of the card over the jar.

Results

The card remains over the mouth of the jar through the entire rotation. The part of the card over the mouth of the jar has no noticeable change in its slight concave (curved inward) shape throughout the rotation.

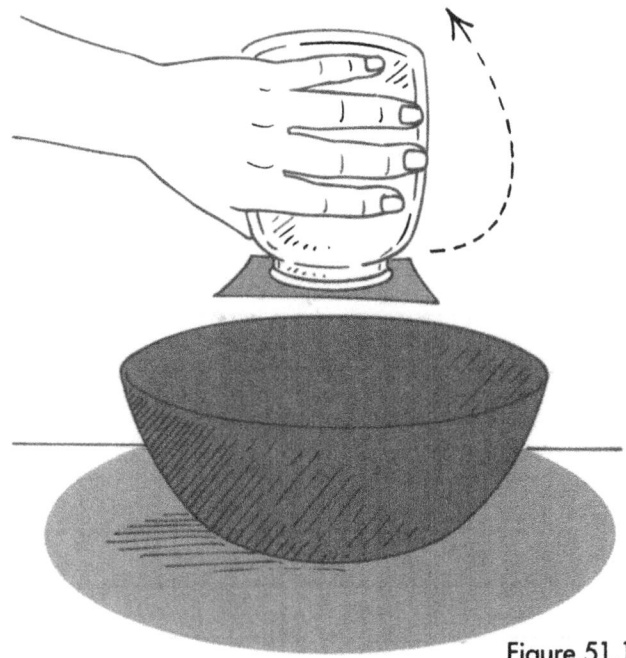

Figure 51.1

Why?

The water around the mouth of the jar wets the paper, and **cohesion** (the force of attraction between like molecules) between the water molecules and **adhesion** (the force of attraction between unlike molecules) between the water molecules and the paper molecules forms a seal. But this seal alone is not enough to overcome the force of gravity. The **concave** (inward curve like the surface of a plate) shape of the card over the mouth of the jar indicates that the card is being pushed into the jar by an outside pressure. This is **atmospheric pressure** (force that gases in the atmosphere exert on a particular area).

At a given point in a fluid, such as air (the name for the mixture of gases in earth's atmosphere), pressure exerted by the fluid acts at right angles at every point on a submerged object. The card stays in place over the mouth of the jar regardless of orientation, showing that an equal air pressure is exerted on the card from all directions. In the same way, at a specific height above Earth, atmospheric pressure is the same on all sides of an object as indicated by no noticeable change in the concave shape of the paper covering.

Try New Approaches

1. Does the size of the jar affect the results? Repeat the investigation using jars of different sizes, but with the same mouth size as the original jar.

2. Does the size of the mouth of the jar affect the results? Repeat the investigation using jars with the same size but with different mouth sizes.

Design Your Own Experiment

1. The relationship between the pressure of a liquid and the depth (height) of that liquid is expressed by the formula $P_{liquid} = dgh$, where d is the density of the liquid, g is the **acceleration** (an increase in velocity) due to gravity (9.8 m/s^2), and h is the height of the fluid column above the point in question. For a specific fluid such as water, if the density of water ($1 \times 10^3 \text{ kg/m}^3$) remains the same throughout, since gravity is relatively constant, the only variable is depth. Thus the pressure on the liquid is directly related to the depth or height of the liquid in the cup.

   Design a way to test this. One way is to fill a container with water and make holes of identical size in the container at various depths (see Figure 33.2). The length (L) of the stream of water spurting from each hole can be used to compare pressure. The greater the depth of the liquid, the greater the pressure. Use a tall paper cup and make a hole near the bottom with the point of a pencil. Put a piece of tape over the hole and fill the cup. Measure the height from the hole to the top of the water's surface. Record this as the depth of hole A in a Pressure Data table like Table 51.1. Use this height and the equation $P_{water} = dgh$ to determine the pressure of water with a density of $1 \times 10^3 \text{ kg/m}^3$ at the depth of hole A measured in meters. Example: The pressure of water at a depth (h) of 0.03 m is:

   $$P_{water} = dgh$$
   $$P_{water} = 1 \times 10^3 \text{ kg/m}^3 \times 9.8 \text{ m/s}^2 \times 0.03 \text{ m}$$
   $$= 0.294 \text{ kg·m·m/m}^3\text{·s}^2$$
   $$= 0.294 \times 10^3 \text{ N/m}^2$$
   $$= 0.294 \text{ kPa}$$

   Note that the units $kg·m·m/m^3·s^2$ can be grouped forming $kg·m/s^2·m^2$.

Since $1 \text{ kg·m/s}^2 = 1 \text{ N}$, then $0.294 \text{ kg·m·m/m}^3\text{·s}^2 = 0.294 \times 10^3 \text{ N/m}^2$, and since $1000 \text{ N/m}^2 = 1 \text{ kPa}$, then $0.294 \times 10^3 \text{ N/m}^2 = 0.294 \text{ kPa}$. Kilopascal (kPa) is a practical metric unit for measuring pressure since Pascal (Pa) is generally too small.

Elevate the cup on an inverted rectangular container at one end of a tray so that the hole points toward the tray. Remove the tape, and mark where the water squirting out of the hole first lands. Measure the distance from the hole to this mark and record it as the length of the water stream for hole A. Repeat this procedure for two other holes, B and C, above hole A, making sure you open only one hole at a time and that all the holes are of equal size.

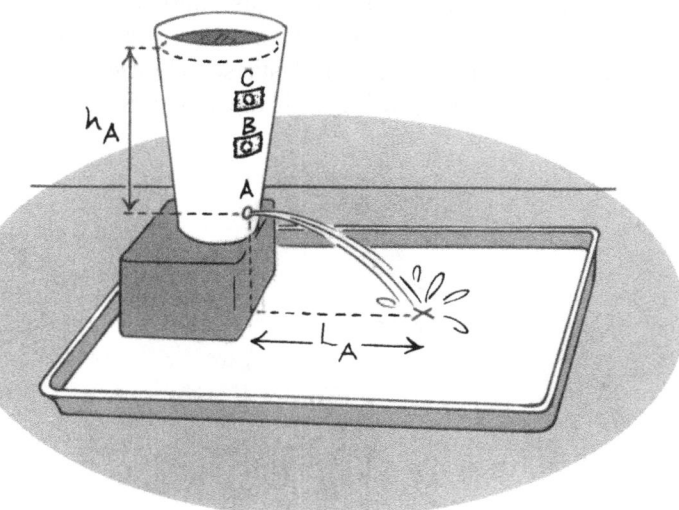

Figure 51.2

	h = Depth of Hole (m)	Pressure (kPa)	L = Length of Water Stream (m)
hole A (bottom)			
hole B (middle)			
hole C (top)			

TABLE 51.1 PRESSURE DATA

2. Blaise Pascal (1623–1662), a French mathematician and inventor, was the first to discover that fluids at rest exert pressure equally in all directions. **Pascal's law** states that the pressure applied to an enclosed fluid is transmitted equally in all directions and to all parts of the enclosing vessel, if the

fluid is incompressible. Design an experiment to test Pascal's law. One way is to squeeze a 2-liter plastic soda bottle containing both water and a transparent condiment with an air bubble. The bubble of air in the condiment will increase or decrease depending on the pressure applied to it. When the bottle is squeezed, the liquid is not compressed but transmits the pressure in all directions, resulting in the compression of the air bubble inside the packet; thus the average density of the packet increases and the packet is less buoyant. Prepare the bottle by first selecting the best condiment packet. Do this by filling a quart (liter) jar about three-fourths full of water and dropping several condiment packets into the water. Select the packet that just barely sinks below the water's surface. Insert the condiment packet into an empty 2-liter plastic soda bottle. Fill the bottle to overflowing with tap water. Secure

Figure 51.3

the cap on the bottle. Then squeeze the bottle with your hands. The condiment packet will sink when the bottle is squeezed and rise when the bottle is released. Note the size of the air bubble in the condiment as it sinks and rises.

3. Earth's atmosphere is kept around Earth by gravity. Atmospheric pressure, which is created by air molecules hitting and bouncing off surfaces, including one another, keeps gravity from pulling all of the air molecules to Earth's surface. Gravity pulls the atmosphere downward and air pressure pushes the atmosphere upward. The result of the original experiment showed equal air pressure from all directions. Design another experiment to measure atmospheric pressure from different directions. One way is to use a **manometer** (an instrument used to measure the pressure of fluids). See Appendix 7 for instructions on how to make a manometer. Test atmospheric pressure by holding one end of the manometer tube in different directions: right, left, up, and down. The water level in the tubes will be even if the pressure on both sides is equal.

Get the Facts

1. In physics, pressure is defined as a force measured over an area. In the International System of Units (SI), one unit of pressure is newtons (N) per square meter. In honor of Pascal, the Pascal (Pa) unit is also used to measure pressure. It is equal to one newton per square meter. There are other units of measuring pressure, such as atmosphere, millimeters of mercury, bar, milibar, barye, and torr. Find out more about Pascals and other units of pressure. How do the other units compare to the Pascal unit? For information see a physics text.

2. How does Pascal's law explain the workings of a hydraulic jack? For information see Karl F. Kuhn's *Basic Physics* (New York: Wiley, 1996), pp. 75–76.

Notes on
Static Fluids

Key Facts:

_____

_____

_____

_____

My Results:

_____

_____

_____

_____

_____

_____

_____

Conclusions:

_____

_____

_____

_____

_____

Results of Try New Approaches:

_____

_____

_____

_____

_____

Notes on Designing My Own Experiment:

_____
_____
_____
_____
_____
_____
_____
_____
_____
_____
_____
_____
_____
_____
_____
_____
_____
_____
_____
_____
_____
_____

Notes on Get the Facts:

_____
_____
_____
_____
_____
_____

52 Friction
Force That Resists Motion

Friction is a force that resists motion and occurs whenever anything moves while in contact with anything else. Everyday actions such as sweeping, brushing your teeth, and turning a doorknob all involve friction. Without friction you couldn't hold on to the broom, toothbrush, or doorknob. Your hand would slip and slide the way your shoes do on ice. Friction also results in the slowing of moving objects. A thrown baseball or Frisbee rubs against air as it moves, and the friction of this contact slows its motion.

In this project, you will measure the static friction (force needed to move an object) of an object. You will discover the difference between an object's static friction and sliding friction (force needed to keep an object moving at a uniform speed). And you will discover which of these types of friction is greater. You will also calculate different types of coefficient of friction (ratio between the force of friction and the weight of the object being moved).

Getting Started

Purpose: To measure the static friction of an object.

Materials

6-by-12-inch (15-by-30-cm) piece of cardboard
transparent tape
9-ounce (270-ml) plastic cup
16-inch (40-cm) piece of string
rubber band (a #1, medium-size, band works well)
scissors
metric graph paper with 1-cm squares
20 marbles
pencil

Procedure

1. Place the cardboard on a table.

2. Tape the cup to the center of one of the shorter sides of the cardboard near the edge.

3. Tie the string around the bottom of the cup. Then tie one of the free ends of the string to the

rubber band. You want the rubber band to be as close to the cup as possible. Cut off the excess ends of the string.

4. Lay the graph paper on the cardboard, with one edge next to the cup. Secure the paper to the cardboard with tape.

5. Add the marbles to the cup.

6. With the point of the pencil inside the loop of the rubber band, pull the rubber band straight but do not stretch it.

7. With the rubber band in this position, make a mark on the graph paper. Then move the pencil in a straight path away from the cup, stretching the rubber band and tracing a line on the graph paper with the pencil point (see Figure 52.1).

8. Stop moving the pencil as soon as the cardboard moves forward.

Length of Line = 19 squares

Figure 52.1

9. Count the squares that your pencil crossed. For partial squares, estimate the fractions. Record the number of squares in a Friction Data table like Table 52.1.

TABLE 52.1 FRICTION DATA						
Surface/ Marbles in Cup	Squares					
	Trial 1	Trial 2	Trial 3	Trial 4	Trial 5	Average
table/20						

10. Repeat steps 7 through 9 four times. Average the results of all five tests, and record the average in the Friction Data table.

Results

You will be able to stretch the rubber band some distance before the cardboard moves. The exact results will depend on the elasticity of the rubber band as well as the surface of the table and the weight of the cup.

Why?

Friction is the name of forces that oppose the motion of one surface relative to another when the two surfaces are in contact with each other. Friction acts parallel to the surfaces in contact and in the opposite direction of motion. Friction is due to the fact that no matter how smooth the **macroscopic** (large enough to be seen with the naked eye) view of a surface, its **microscopic** (so small it requires a microscope to be seen) view is rough. The irregularities on both surfaces that are rubbing against each other interlock and offer resistance to motion. **Static friction** is the force that opposes the start of motion of an object.

In this investigation, the distance the rubber band stretches (length of line in number of squares) indicates the force of static friction between the surface of the cardboard and the surface of the table. The more the rubber band stretches before the cardboard moves, indicated by the number of squares crossed by the line, the greater the force of static friction.

Try New Approaches

1. How does the smoothness of the surfaces affect static friction? Repeat the investigation, placing the cardboard on different surfaces, such as waxed paper and different grades of sandpaper, that have been secured to the table.

2. How does the force pressing the surfaces together affect static friction? Repeat the original investigation, increasing the weight on the cardboard by adding more marbles to the cup.

3. How does lubrication affect static friction? Repeat the investigation twice: First use a surface of sandpaper alone. Then use sandpaper covered with a thick layer of petroleum jelly.

Design Your Own Experiment

1a. Design a way to measure the static friction of a system. One way is to attach one end of a string to a small box and the other end to a paper cup. Set another cup filled with marbles in the box, and place the box on a table so that the empty paper cup hangs over the edge of the table. The box should start out about 6 inches (15 cm) from the edge of the table. Add weights to the empty hanging cup until the box system (box and contents) starts to move. This weight equals the static friction (F_f) of the system. For weights, use something like coins, paper clips, and/or washers that you know the individual weight of. Ask your teacher or a pharmacist to weigh whichever materials you choose on an accurate scale.

b. Friction between any two surfaces can be measured by the coefficient of static friction, which is the ratio between the force of static friction between two surfaces in contact with each other and the force holding the surfaces together. The coefficient of static friction is a constant that depends on the nature of the surfaces in contact with each other. Determine the coefficient of static friction between the cardboard and the material on the table's surface using this equation:

$$\mu = F_f / F_N$$

In this equation, μ (mu) is the symbol for the coefficient of static friction when F_f is the static friction (the total weight in the hanging cup) when the box moves, and F_N is the perpendicular force pushing the surfaces together (on a horizontal surface, the weight of the box and its contents). For information about the coefficient of static friction for common surfaces, see a physics text.

2a. **Sliding friction** is the frictional force between objects that are sliding with respect to one another. Using the materials from the previous investigation, start by removing about half of the weights from the hanging cup. Then slowly add the weights to the hanging cup one at a time, but this time after each addition, give the box a slight push toward the edge of the table where the cup is hanging. Continue this process until the box starts to move at a uniform **velocity** (speed and direction of a moving object). (If the box **accelerates**—changes in velocity per time—the force is too large; if the box stops, the force is too small.) How does the sliding friction compare to the static friction of the box?

b. How does the contact area of surfaces affect sliding friction? Design a way to measure the sliding friction of a system in which only the surface area changes. One way is to replace the box with a block of wood with different-size faces. Then determine the sliding friction of the wood for each of its different-size faces.

c. Determine the **coefficient of sliding friction** (the ratio between the force of sliding friction between surfaces in contact with each other and the force holding the surfaces together) for the box and table surfaces using the equation $\mu = F_f / F_N$, where μ (mu) is the symbol for sliding friction when F_f is the sliding friction.

cup with marbles inside

cup with weights inside

Figure 52.2

Get the Facts

1. You can move one object across another without sliding, and thus without sliding friction, using rollers. What are ball bearings and, when placed between surfaces, how do ball bearings minimize sliding friction? For information, see Louis A. Bloomfield, *How Things Work: The Physics of Everyday Life* (New York: Wiley, 1997), pp. 57–60.

2. Brakes on a car work because of friction. Why are brakes generally designed to be applied to the front wheels before the rear wheels? For an exploratory investigation to discover the answer to this question, see Robert Gardner, *Experiments with Motion* (Springfield, N.J.: Enslow Publishers, 1995), pp. 63–67.

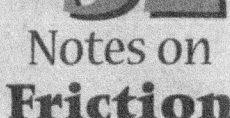

Notes on
Friction

Key Facts:

_____
_____
_____
_____

My Results:

_____
_____
_____
_____
_____
_____
_____

Conclusions:

_____
_____
_____
_____
_____

Results of Try New Approaches:

_____
_____
_____
_____
_____

Notes on Designing My Own Experiment:

_____
_____
_____
_____
_____
_____
_____
_____
_____
_____
_____
_____
_____
_____
_____
_____
_____
_____
_____
_____
_____
_____
_____
_____
_____
_____
_____

Notes on Get the Facts:

_____
_____
_____
_____
_____
_____

53 Work
Force through a Distance

"Work" is a term used in physics to indicate that a force has caused an object to move. A tennis player hitting a tennis ball with a racket is an example of work because the tennis ball moves when struck by the racket. But work is not done every time a force is applied. For example, if you push against a building for several minutes, you may become tired, but no work has been done. This is because the building did not move. The amount of work done is determined by multiplying the force applied to an object by the distance the object moves in the direction of the force. The SI (internationally agreed-upon method of using the metric system of measurement) unit for work is joule if the SI units for force and distance, respectively, are newton and meter.

In this project, you will learn how to measure the work done on an object. You will determine the effect that a simple machine has on work. You will also determine the effect of the direction of the force on work.

Getting Started

Purpose: To measure the work done on an object.

Materials

brick (or any object of comparable weight)
shoe box large enough to hold the brick
paper hole-punch
5-pound (2200-g) spring scale with a hook
yardstick (meterstick)
masking tape

Procedure

1. Place the brick inside the box.

2. Use the paper hole-punch to cut a hole in the end of the box. Attach the scale's hook through the hole in the box (see Figure 53.1).

3. Place the box at one end of a table. Place a piece of masking tape 6 inches (15 cm) in front of the box. This will be the starting line.

4. Measure the distance from the starting line to the end of the table toward which you are pulling in **meters (m)**—an SI unit for distance. Record this as distance traveled *(d)* in a Work Data table, like Table 53.1.

5. Pull the scale so that the scale and box move horizontally across the table at a constant speed. Determine the scale reading when the box crosses the start line. This force must be measured in **newtons (N)**—an SI unit for force. Record the newton force in the Work Data table. If your scale measures in pounds, convert pounds to newtons using this conversion: 1 pound = 4.45 N. For example, if the measurement is 2 pounds, the newton force would be 2 pounds × 4.45 N/1 pound = 8.9 N. If the scale measures in grams, convert grams to newtons using this conversion: 1 g = 0.0098 N. For example, if the measurement is 908 g, the newton force would be 908 g × 0.0098 N/1g = 8.9 N.

6. Repeat step 5 four times and average the measurements.

7. Calculate the work done on the box using this equation: $w = f \times d$.

In the equation, w equals work, f is the force acting on the box in newtons, and d is the distance (in meters) the box moves during the time the force is being measured. For example, if $f = 8.9$ N and $d = 0.5$ m, then $w = 8.9$ N × 0.5 m = 4.45 Nm.

Figure 53.1

TABLE 53.1 WORK DATA

Distance (d), m	Force (f), N						Work (w), J
	Trial 1	Trial 2	Trial 3	Trial 4	Trial 5	Average	$w = f \times d$

Note: 1 Nm = 1 joule (J). **Joule (J)** is an SI unit for work.

So for this example the work done is 4.45 J.

Results

The work done will depend on the force needed to pull the box and the distance it moves. In the example, the work done is 4.45 J.

Why?

Work is what is accomplished when a force causes an object to move. The amount of work done is equal to the product of the force applied to an object times the distance the object moves in the direction of the force. Another requirement for work to be done is that the distance the object is moved must be in the same direction that the force is applied. In this experiment, a horizontal force moves the box in a horizontal direction, so work is done.

Try New Approaches

1. Does the speed at which an object moves affect the work needed to move it? Repeat the experiment twice, first at a higher but constant speed and then at a lower but constant speed.

2. How does the weight of the object being moved affect the work done to move it? Repeat the original experiment twice, first using a lesser weight in the box and then using a greater weight. *Note:* Try to pull the box at the same speed for each testing.

Design Your Own Experiment

1. A **machine** is a device that makes work easier. Machines make work easier by changing either the size or the direction of the input force. **Simple machines** are the most basic machines, such as an **inclined plane** (a flat, slanted surface). Inclined planes are used to transport an object to a specific height. Design an experiment to determine if using an inclined plane affects the overall work done on the object being moved. One way is to add weight, such as marbles, clay, or coins, to a small box with a lid. Close the box and secure the lid with tape. Tie a string around the box and attach the hook of a spring scale to the string. Use the scale to slowly raise the box a vertical distance of 1 meter. As you raise the box, ask a helper to note the reading on the scale in newtons, grams, or pounds. If the reading moves up and down slightly, record the average reading. Employ the previous method of determining force in newtons using pound or gram units. Then determine the work done in lifting the box using this equation: $w = f \times d$. Then prepare an inclined plane by placing one end of a board at least 1 meter longer than the box on a stack of several books. Use the scale to move the box up the inclined plane for a distance of 1 meter. Repeat the procedure for determining the force needed to move the box and the work done. Use diagrams to display the results of the experiments.

2a. Sometimes a force on an object is at an angle to the direction of motion. An example would be pulling a wagon's handle at an angle, causing the wagon to move horizontally (see Figure 53.2 on the next page). In this case, the relationship of the force acting on the wagon can be expressed by the equation $d_a/d_h = f_h/f_a$, where d_a is the distance of the side adjacent to the angle of the applied force, d_h is the distance of the hypotenuse (side opposite the right angle), f_h is the force causing horizontal motion parallel to the direction in which an object is moved, and f_a is the force applied at angle A°. The **cosine (cos)** of an angle is equal to the length of the adjacent side (d_a) divided by the hypotenuse (d_b). Since $\cos A° = d_a/d_h$ and $d_a/d_h = f_h/f_a$, then

Figure 53.2

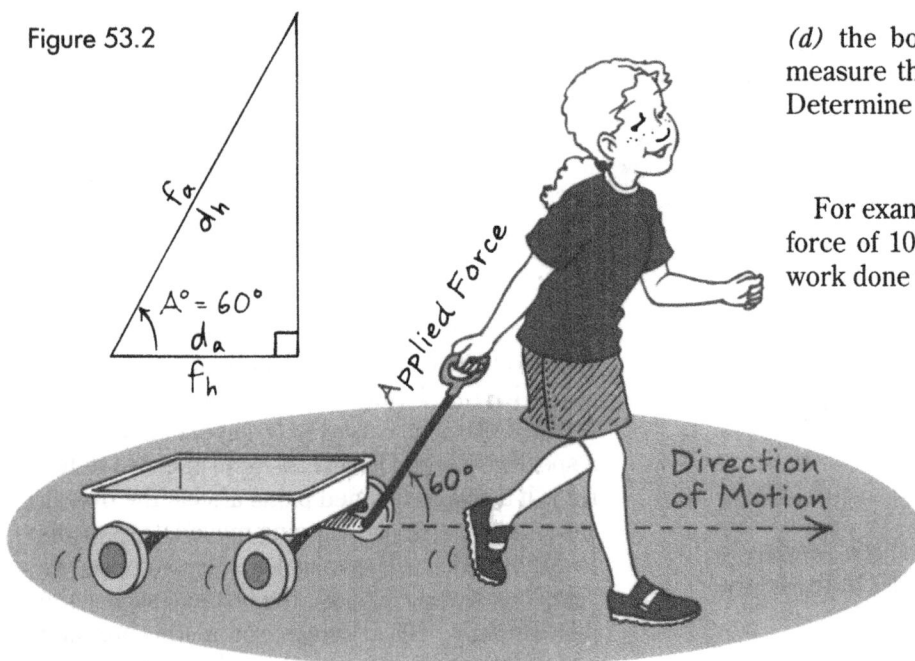

$\cos A° = f_h/f_a$. Thus the horizontal force (f_h) causing the wagon to move in a horizontal direction can be calculated using this equation: $f_h = f_a \times \cos A°$. (See Appendix 8 for the cosine value of different angles.)

Design an experiment to calculate the work done by a force that is at an angle to the direction in which an object is moved. One way is to attach a scale to a weighted box. Move the box across a table by pulling on the scale so that this force is at an angle to the movement of the box, as shown in Figure 53.3. Measure and record the distance

(d) the box is moved. Use a protractor to measure the angle (A°) of the applied force. Determine the work using this equation:

$$w = (f_a \cos A°) \times d$$

For example, if the box is moved 0.6 m by a force of 10 N applied at an angle of 30°, the work done would be:

$$w = (10 \text{ N} \times \cos 30°) \times 0.6 \text{ m}$$
$$= 10 \text{ N} \times 0.87 \times 0.6 \text{m}$$
$$= 5.22 \text{ Nm or } 5.22 \text{ J}$$

For more information about work done by a constant force that is applied at an angle relative to the direction of motion, see J. P. Den Martog, *Mechanics* (New York: Dover, 1961), pp. 133–135.

b. How does the angle affect the amount of work done in the previous experiment? Repeat the experiment three times, first at a smaller angle and second at a greater angle, but less than 90°. For the third trial, use an angle of 90°, thus slightly lifting the box above the table. Prove mathematically that while the box is moved horizontally while applying a force at 90°, no work is done. **Science Fair Hint:** Show vector diagrams for each angle. You do work in lifting an object, but once the object is lifted, you do no work in carrying it across a room. For an explanation of this seeming paradox, see *work* in a physics text and Larry Gonick and Art Huffman, *The Cartoon Guide to Physics* (New York: HarperPerennial, 1990), p. 75.

Get the Facts

Power is the rate of doing work. Since power is work divided by time, power is expressed as joules per second in SI units. The power unit of watt was named after James Watt (1736–1819), the inventor of the steam engine. How do the units of watt and horsepower compare to the SI unit of joules/sec? See a physics text for a comparison of power units.

Figure 53.3

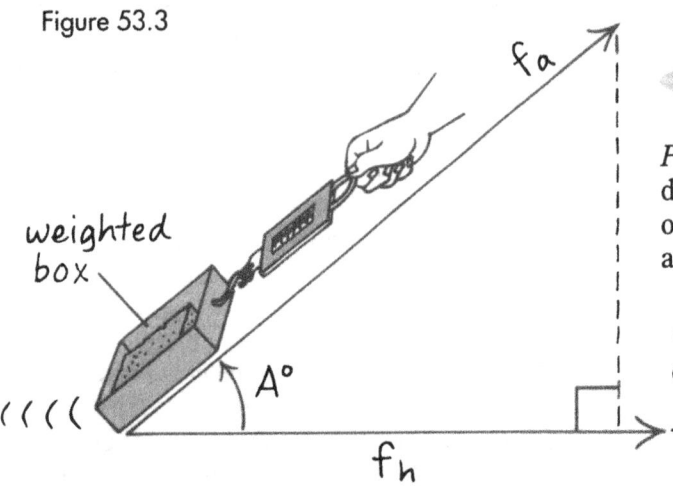

Notes on
Work

Key Facts:

_____
_____
_____
_____

My Results:

_____
_____
_____
_____
_____
_____
_____
_____

Conclusions:

_____
_____
_____
_____
_____

Results of Try New Approaches:

_____
_____
_____
_____
_____

Notes on Designing My Own Experiment:

_____
_____
_____
_____
_____
_____
_____
_____
_____
_____
_____
_____
_____
_____
_____
_____
_____
_____
_____
_____
_____
_____

Notes on Get the Facts:

_____
_____
_____
_____
_____
_____

54. Newton's Third Law of Motion
Action-Reaction

Isaac Newton (1642–1727), the famous British scientist credited with discovering gravity, also gave us three laws describing motion. Newton's first law of motion states that a force is needed to change the motion of an object. In other words, a force either starts an object moving or causes a moving object to stop. His second law of motion explains how the force needed to accelerate (a change in velocity) an object depends on the mass of the object. His third law explains that forces act in pairs.

In this project, you will demonstrate Newton's third law of motion, that every action has an equal and opposite reaction due to the action of paired forces. You will also determine how pairs of forces that are equal but acting in opposite directions can produce motion.

Getting Started

Purpose: To demonstrate Newton's third law of motion.

Materials

pencil
5-ounce (150-ml) paper cup
40 to 50 pennies
12-inch (30-cm) piece of string
1 handheld spring scale

Procedure

1. Use the pencil to make two holes across from each other just beneath the rim of the cup. Place the coins in the cup.

2. Loop the string through the holes, then tie the ends of the between the holes.

3. Hold the scale and adjust it so that it reads zero.

4. While holding the scale, attach the cup so that the cup hangs freely. Observe the reading on the scale.

Results

The cup pulls the scale down so that the measurement on the scale indicates the weight of the cup and the coins.

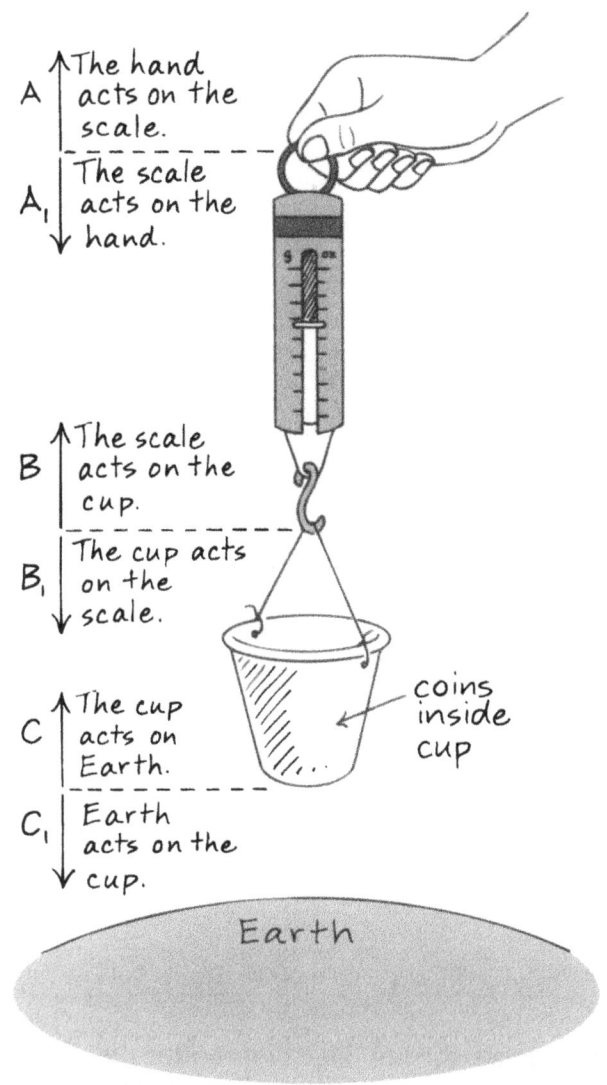

Figure 54.1

Why?

Newton's third law of motion states that for every action there is an equal and opposite reaction due to pairs of forces. In other words, Newton realized that if one object exerts a force on another, the second object exerts an equal force but in the opposite direction on the first object. You can be sure that two forces are action-reaction pairs of forces if the reverse description of one force describes the other force. In Figure 54.1, the three identified action-reaction pairs of forces are A/A_1; B/B_1; C/C_1. The description of force A is "the hand acts on the scale," and the description of force A_1 is "the scale acts on the hand." One description is the reverse of the other, so the forces are equal in magnitude, opposite in direction, and act on different objects. Thus forces A and A_1 are an action-reaction pair.

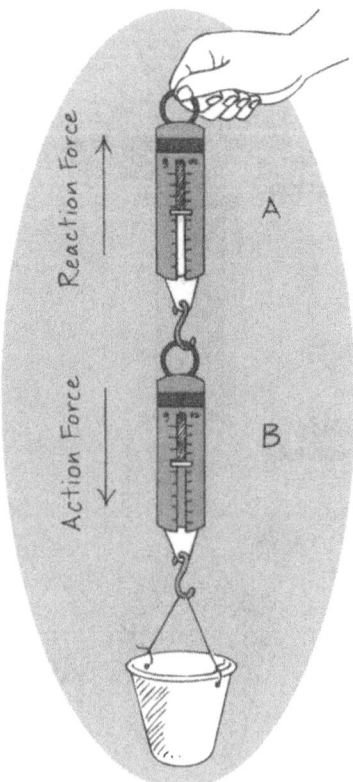

Figure 54.2

Try New Approaches

The scale attached to the cup measures the downward force of the cup (the action). How can the upward force of the hand be measured (the reaction)? Repeat the experiment using two scales. First hang one scale from the other. So that the weights of the scales are not considered, while holding the top scale (called A), adjust the scales so each reads zero. Let the top scale be A and the bottom scale be B. Attach the cup to the bottom scale (called B) as before. The reading on scale A measures the upward force (the reaction force) and scale B measures the downward force (the action force). **Science Fair Hint:** A diagram showing the action-reaction pairs can be used as part of a project display.

Design your Own Experiment

1. Every force on an object causes the object to be compressed to some degree. Design an experiment to demonstrate that an object compresses until the action-reaction forces are equal. For example, fill a 3-ounce (90-ml) cup with coins. Lay two similar-size books about 10 inches (25 cm) apart on a table. Support the ends of a thin, flexible, plastic ruler on the books. Set the cup of coins in the center of the ruler. **Science Fair Hint:** Make a diagram showing the compression of the ruler, the action-reaction pairs of forces, such as in Figure 54.3. Three legends describing the force pairs of the books and the cup can be added to the drawing. For example:

Force Pairs for Book A

- Ruler acts on book A.
  Book A acts on ruler.

- Book A acts on table.
  Table acts on book A.

- Table acts on Earth.
  Earth acts on table.

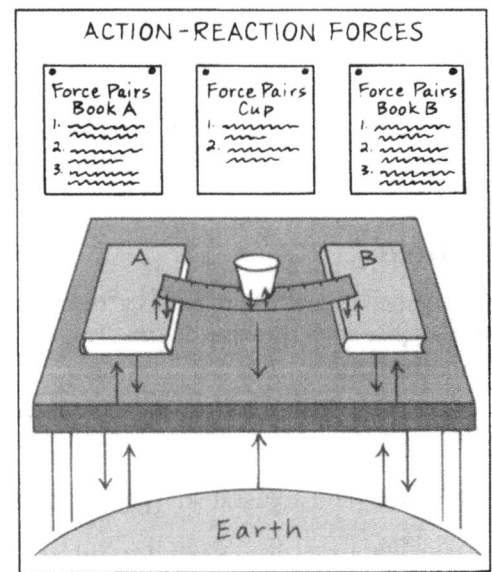

Figure 54.3

2. Unaccompanied forces do not exist. Since all forces are in pairs of equal strength, acting in

opposite directions, and acting on different objects, what causes motion? A **resultant force,** is the single force that has the same effect as the sum of two or more forces acting simultaneously on one object. When forces simultaneously act on an object and the resultant force is zero, the forces are said to be **balanced forces** and produce no acceleration. When the resultant force of a group of forces acting simultaneously on an object is not equal to zero, the forces are said to be **unbalanced forces. Newton's first law of motion** explains that an unbalanced force acting on an object is needed to cause acceleration. In this law, the resultant force is the **net force** (the sum of all forces acting simultaneously on an object).

Design an experiment to demonstrate that a pair of action-reaction forces are unbalanced because they act on different objects. One way is with two identical balloons. Inflate one of the balloons and tie a knot in its open end. Lay the balloon on a table and observe any motion of the balloon. **Science Fair Hint:** Prepare a diagram representing the action-reaction forces for the open and closed balloons, such as in Figure 54.4. Write the calculations for determining the net force of the gas inside the closed balloon on the balloon, represented by forces A as well as the net force of the balloon on the gas inside the balloon, represented by forces B. The equation for net force is:

$$F_{(net\ force)} = (F\uparrow + F\downarrow) + (F\rightarrow + F\leftarrow)$$

The calculation for determining the net force of the gas on the balloon in the closed balloon is:

$$
\begin{aligned}
F_{(net\ force)} &= (F\uparrow + F\downarrow) + (F\rightarrow + F\leftarrow) \\
&= (A_1\uparrow + A_3\downarrow) + (A_2\rightarrow + A_4\leftarrow) \\
&= 0 + 0 \\
&= 0
\end{aligned}
$$

The equation representing the net forces of the gas acting on the open balloon is:

$$
\begin{aligned}
F_{(net\ force)} &= (F\uparrow + F\downarrow) + (F\rightarrow + F\leftarrow) \\
&= (A_1\uparrow) + (A_2\rightarrow + A_3\leftarrow) \\
&= A_1\uparrow + 0 \\
&= A_1\uparrow
\end{aligned}
$$

In the closed balloon, the net force is zero, therefore there is no unbalanced force and thus no motion. In the open balloon, the net force is equal to force $A_1\uparrow$. Thus there is an unbalanced force in the up direction causing the balloon to move upward. Use the net force equation to calculate

Legend	
Forces	Description
A	Gas inside balloon is acting on balloon.
B	Balloon is acting on gas inside balloon.

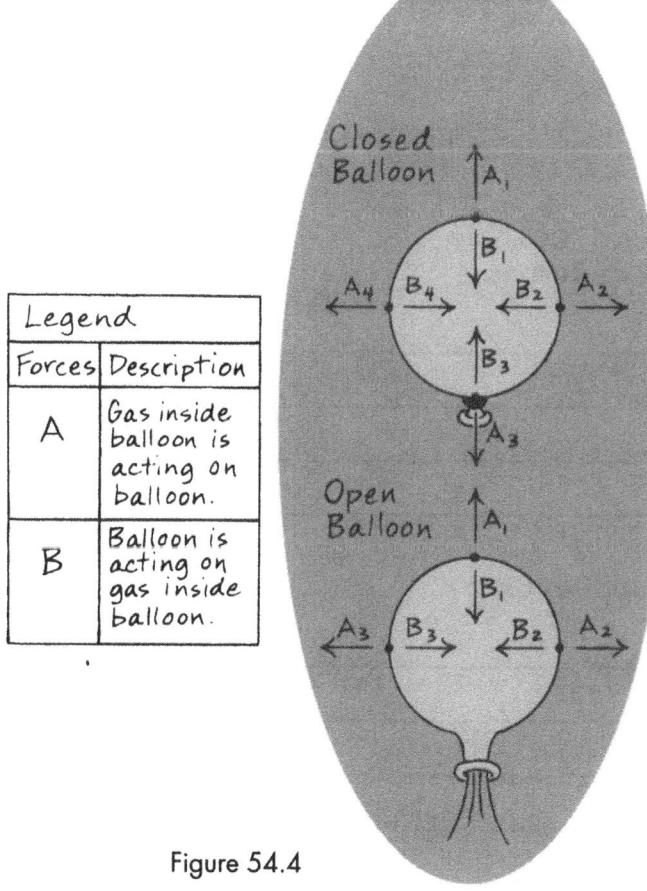

Figure 54.4

the net force of the balloon on the gas inside the open and closed balloons to determine why the gas moves out of the open balloon.

Get the Facts

1. The use of steam as a source of power can be traced back to a toy invented by a Greek engineer named Hero of Alexandria (20?–62?). This toy turned as a result of action-reaction forces. For information about the construction of Hero's toy and how the action-reaction forces were produced, see Struan Reid's and Patricia Fara's *Inventors* (Tulsa, Okla: EDC, 1994), p. 10.

2. A simple Hero's engine can be made from a soda can. For information about building this simple Hero's engine and making predictions about its movement see Robert Ehrlich's *Why Toast Lands Jelly-Side Down* (Princeton, N.J.: Princeton University Press, 1997), pp. 69–71.

Notes on
Newton's Third Law of Motion

Key Facts:

_____
_____
_____
_____

My Results:

_____
_____
_____
_____
_____
_____
_____
_____
_____

Conclusions:

_____
_____
_____
_____
_____

Results of Try New Approaches:

_____
_____
_____
_____
_____

Notes on Designing My Own Experiment:

_____
_____
_____
_____
_____
_____
_____
_____
_____
_____
_____
_____
_____
_____
_____
_____
_____
_____
_____
_____
_____
_____
_____
_____

Notes on Get the Facts:

_____
_____
_____
_____
_____
_____

55 Polarization
Vibrations in One Direction

Unpolarized light consists of waves with electric fields vibrating in all directions. In some cases, however, all the waves in a beam vibrate in one direction or in one plane. Such light is said to be polarized.

In this project, you will investigate polarized and unpolarized light. You will determine what effect an analyzer has on polarized light. You will determine if the angle of incidence (angle at which light strikes a surface) affects the degree of horizontal polarization. You will also test the optical activity of different solid materials and of different concentrations of water solutions.

Getting Started

Purpose: To polarize light.

Materials

desk lamp with incandescent bulb
inexpensive plastic polarized sunglasses

Procedure

1. Turn on the lamp and position it so that the bulb is visible to you. Stand at a distance of about 3 feet (1 m) from the bulb. Look at the bulb and make note of its brightness.

2. Remove the lenses from the sunglasses by twisting the frames and popping the plastic lenses out.

3. At a distance of about 3 feet (1 m), close one eye and look through one of the polarized lenses at the lit bulb and again note the bulb's brightness. This will be called lens A.

4. Hold the second lens (called lens B) in front of but not touching lens A and while still closing one eye look at the light through both lenses. Hold lens A in place while rotating lens B until the bulb appears at its brightest as viewed through both lenses. Then slowly rotate lens B 90°, observing any change in the brightness of the bulb.

Figure 55.1

Results

The bulb is less bright when viewed through one lens than with the naked eye. Viewing the bulb through two lenses further decreases its brightness. As one of the lenses is rotated in front of the other lens, the light decreases still further until it is no longer visible or only partly visible.

Why?

Visible light is a form of radiation, which is energy that travels in the form of **electromagnetic waves** (transverse waves consisting of an electric field and a magnetic field vibrating at right angles to each other and to the direction of the propagation of the wave). A **transverse wave** is one in which the vibrations are perpendicular to the direction in which the waves are traveling, like the up-and-down or vertical motion of water waves. But unlike water waves, light waves can vibrate in all directions perpendicular to the direction of motion. **Polarized light** is light in which the electric fields of the light waves vibrate in a direction parallel to each other. A light wave whose electric field is vibrating in the vertical direction is said to be

vertically polarized. A light wave whose electric field is vibrating in the horizontal direction is said to be **horizontally polarized. Unpolarized light** contains light waves with electric fields vibrating in directions that are not parallel with one another, such as the light from the bulb in this experiment. (See Figure 55.2.) A polarized lens acts as a **polarizer,** which is a material that allows electric fields of light vibrating in only one direction to pass through it. When unpolarized light strikes a polarized lens, part of the light is **reflected** (bounced off of), part is absorbed by the lens, and part of the light with electric fields vibrating in one specific plane passes through. The light that emerges from the other side of the polarizer contains light waves with electric fields vibrating only in one direction or in one plane and is said to be **polarized** (also called **plane polarized**). Figure 55.2 represents vertical polarization. Polarized sunglasses are generally made of plastic material in which needlelike microscopic crystals are embedded. These crystals line up parallel to one another and make the polarized lens act as though it consists of many slitlike openings parallel to one another. So only those light waves with electric fields vibrating in the same plane as the parallel slits in the polarizer can get through.

Placing the two lenses together demonstrates the effect of using two polarizers aligned with each other. The first lens in line with the light is called the polarizer, and the second lens is called the **analyzer** (a polarizer used to determine if light is polarized).

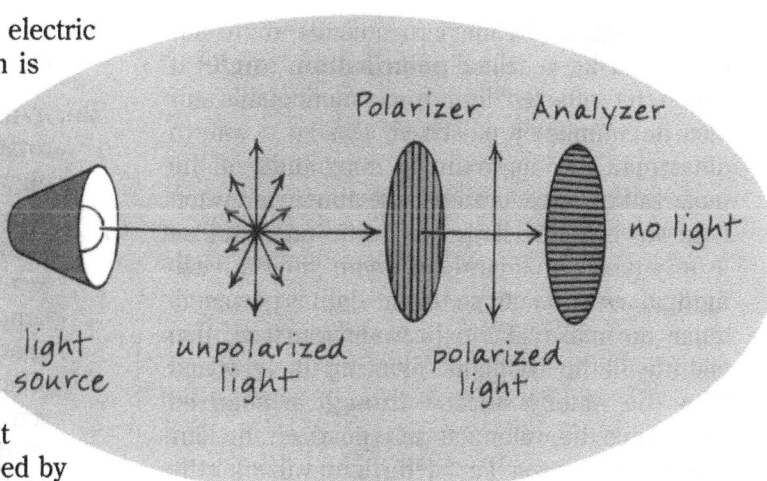

Figure 55.3

When the crystals in the two lenses are lined up parallel to one another, the greatest amount of light possible passes through. In this position, rotating the analyzer 90° results in the crystals in the separate lenses being at right angles to one another. None of the polarized light is able to pass through the analyzer in this position, as shown in Figure 55.3. **Science Fair Hint:** Use diagrams such as the ones shown to represent the results of this investigation.

Try New Approaches

How is light affected by nonpolarized lenses? Repeat the experiment using the lenses from an inexpensive pair of nonpolarized sunglasses.

Design Your Own Experiments

1a. Unpolarized light becomes partially polarized when reflected from nonmetallic surfaces, such as water or glass, while some is transmitted and/or absorbed. The parallel components of the **incident light** (the light striking a surface) is largely reflected. The reflected light is partially polarized in a direction parallel with the surface it reflects from. Thus for lakes and other surfaces parallel to Earth's surface, the reflected light is said to be horizontally polarized. There is a special **angle of incidence** (the angle between the

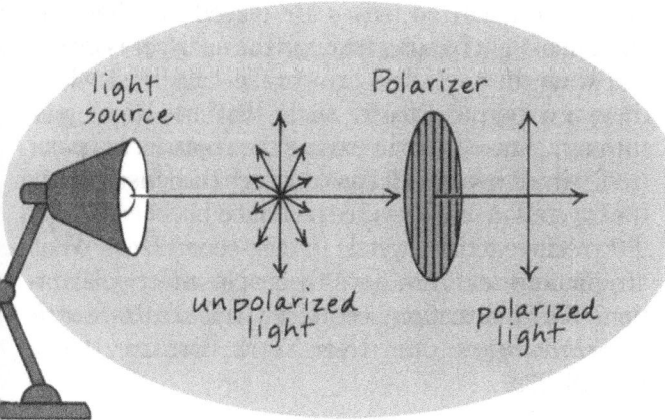

Figure 55.2

incident light and a line perpendicular to the surface it strikes) called **polarization angle** at which the reflected light from a nonmetallic surface is completely polarized. Design a way to determine the approximate magnitude of the polarization angle for light reflecting from water. One way is to fill a large bowl with water and set it on a table. In a darkened room, hold a flashlight above the water so that its light is perpendicular (normal or 90°) to the water's surface, then slightly tilt the flashlight. Sit facing the bowl and view the water's surface through a polarized lens. Since the water acts as a polarizer, the lens acts as an analyzer. Part of the light will enter the water and form a light spot on the bottom of the bowl, and part will reflect, thus forming an **image** (a representation of a physical object formed by light reflected from a surface) of the end of the flashlight. Rotate the analyzer to determine if the reflected light from the water's surface is polarized. The degree of polarization will be determined by how much of the reflected light (the image) disappears. If you see no change in the image, it indicates that the reflected light is not polarized. If all of the image disappears, then the reflected light is 100% polarized. Increase the angle of incidence by tilting the flashlight from its perpendicular position and analyze the reflection at the new angle. Continue this until the flashlight is held parallel to the water's surface. Estimate the angle of the flashlight from perpendicular (the incidence angle) and devise a way to compare the degree of polarization for each angle, such as a rating of 0 to +5, with +5 being the greatest polarization. For information about polarized angles see Craig F. Bohren's *What Light through Yonder Window Breaks?* (New York: Wiley, 1991), pp. 37–40. For information about horizontal and vertical polarization see Louis A. Bloomfield's *How Things Work* (New York: Wiley, 1997), pp. 494, 498, and 529–530.

b. How does the material of the reflective surface affect polarization? Repeat the previous investigation using different materials including a smooth metal, such as a flat baking pan. For information about polarization from reflected surfaces, see of Karl F. Kuhn's *Basic Physics* (New York: Wiley, 1996), pp. 79–80.

2a. A material is said to be **optically active** when it rotates the plane of the electric light waves passing through it. Design a way to test the optical activity of materials, such as by placing a material between two polarized lenses. For example, to test the optical activity of a plastic cup, place the lamp, cup, and analyzer in a straight line with one another. Rotate the analyzer and observe the changing patterns of color. To test the optical activity of transparent tape, you can stretch some of the tape across a small, open frame cut from poster board. Repeat the experiment replacing the cup with the frame of tape. For information about the colors seen in optically active materials when viewed through an analyzer see Hazel Rossotti's *Colour: Why the World Isn't Grey* (Princeton, N.J.: Princeton University Press, 1993), pp. 49–51.

b. Sugar and tartaric acid (cream of tartar) are known to be optically active. Design a method of testing different concentrations of solutions of water and these chemicals. Note that the container used to hold the solutions must not be optically active. Test light passing through empty containers for optical activity and select one that is not optically active. Devise a method of comparing the degree of optical activity of each concentration.

Get the Facts

When two polarized lenses are placed together, the amount of light passing through the analyzer is greatest when their crystals are parallel and least when they are perpendicular, such that no light gets through. Since the light exiting the polarizer is polarized, why does any of it pass through the lenses when the crystals in one lens are positioned between 0° and 90° relative to the crystals in the second lens? What are photons and how does their spin affect polarization? For information see John Gribbin's *In Search of Schrödinger's Cat* (New York: Bantam, 1984), pp. 218–229.

Notes on
Polarization

Key Facts:

_____

_____

_____

_____

My Results:

_____

_____

_____

_____

_____

_____

_____

_____

Conclusions:

_____

_____

_____

_____

Results of Try New Approaches:

_____

_____

_____

_____

Notes on Designing My Own Experiment:

_____
_____
_____
_____
_____
_____
_____
_____
_____
_____
_____
_____
_____
_____
_____
_____
_____
_____
_____
_____
_____
_____
_____
_____
_____

Notes on Get the Facts:

_____
_____
_____
_____
_____
_____

Appendix 1

Random Error of Measurement

Purpose: To determine the uncertainty of experimental measurements.

Materials

calculator

Procedure

1. Calculate the average of the measurements. For example, the average of the same measurements in Table A1.1 is:

   $$\text{average} = (28° + 30° + 31° + 29° + 30°) \div 5$$

   $$= 29.6°$$

2. Calculate the random error (E) of the average measurement using these steps:

TABLE A1.1 ANGLES OF MEASUREMENTS	
Measurement	Angle, °
1	28
2	30
3	31
4	29
5	30

- Calculate the range (R) of the measurements by finding the difference between the largest and smallest measurement values. In Table A1.1, the difference between the largest and smallest measured angle is:

  $$R = 30° - 28°$$

  $$= 2°$$

- Calculate the random error by dividing the range by $\sqrt{(n-1)}$, where n is the number of measurements. The equation that expresses this is:

  $$E = R \div \sqrt{(n-1)}$$

  For the sample data, the random error would be:

  $$E = 2° \div \sqrt{(5-1)}$$

  $$= 2° \div \sqrt{4}$$

  $$= 2° \div 2$$

  $$= 1°$$

- To express the random error of your measurements, you would say that the value of the angle is $29.6° \pm 1°$. This means that the measurement is between 28.6° and 30.6°.

Appendix 2

Relative Error: Percentage Error

Purpose: To calculate the relative error of experimental measurements.

Materials

calculator

Procedure

1. Calculate the average of the measurements. For example, the average of the sample experimental measurements in Table A2.1 is:

$$\text{average} = (28° + 30° + 31° + 29° + 30°) \div 5$$
$$= 29.6°$$

TABLE A2.1 ANGLES OF MEASUREMENTS	
Measurement	Angle, °
1	28
2	30
3	31
4	29
5	30

2. Use the following equation to determine the relative error (also called percentage error) of the measurements. For example, if the accepted (true) value for the angle measurement is 29.8°, the relative error for the experimental measurement would be:

$$E_r = E_a \div A \times 100\%$$

where E_r is the relative error, E_a is the absolute error (the difference between the true and experimental measurements), and A is the true measurement. Note that E_a is the experimental measurement minus the accepted true measurement. For this sample, the absolute error is $29.6° - 29.8°$ $E_r = -0.2°$.

$$E_r = -0.2° \div 29.8° \times 100\%$$
$$= -0.67\%$$

The sign of the absolute or relative error merely indicates whether the result is low (−) or high (+).

Appendix 3

Planet Facts and Figures

Celestial Body	Diameter, miles (km)	Average Density g/ml (water = 1)	Albedo	Aphelion, or Greatest Distance from Sun, millions of miles (millions of km)	Perihelion, or Least Distance from Sun, millions of miles (millions of km)	Average Distance from Sun, millions of miles (millions of km)	Period of Rotation, hours
Mercury	3,047 (4,878)	5.4	0.1	44 (70)	29 (46)	36 (58)	1,407.5
Venus	7,562 (12,100)	5.3	0.76	68 (109)	67 (107)	68 (108)	5,832
Earth	7,973 (12,757)	5.5	0.39	95 (152)	92 (147)	93 (149)	24
Mars	4,247 (6,796)	3.9	0.16	156 (249)	129 (207)	143 (228)	24.6
Jupiter	89,875 (143,800)	1.3	0.52	510 (816)	463 (741)	486 (778)	9.8
Saturn	75,412 (120,660)	0.7	0.61	942 (1,507)	842 (1,347)	892 (1,427)	10.2
Uranus	31,949 (51,118)	1.2	0.35	1,875 (3,000)	1,712 (2,740)	1,794 (2,870)	15.2
Neptune	30,937 (49,500)	1.7	0.35	2,838 (4,540)	2,782 (4,452)	2,810 (4,497)	16
Pluto	1,434 (2,294)	2.0	0.5	4,604 (7,366)	2,771 (4,434)	3,688 (5,900)	153

Appendix 4

Random Sampling in an Open Ecosystem

Use the following procedure to calculate the abundance of plant life in each sampling area.

1. Count the number of each plant type.

2. Determine the total number of all plants.

3. Determine the abundance of each plant type as a percentage of the total number of all the plants.

Example

If there are 30 woody plants and 300 total plants in a sampling area, the abundance of woody plants is calculated as follows:

$$\text{abundance} = \frac{30 \text{ woody plants}}{300 \text{ total plants}} \times 100\%$$

$$= \textbf{10\% woody plants}$$

Use the following procedure to calculate the density of plant life in each sampling area.

1. Count the number of each plant type.

2. Determine the area of the sampling plot.

3. Divide the total number of each plant type by the area of the sampling plot.

Example

If there are ten trees in a 100-square-yard (100-square meter) area, the density of trees is calculated as follows:

$$\text{density} = \frac{10 \text{ trees}}{100 \text{ yd}^2 \text{ (m}^2)} = \textbf{0.1 tree/yd}^2 \textbf{ (m}^2)$$

Use the following procedure to calculate the frequency of each plant type.

1. Determine how many of the subplots have each type of plant.

2. Determine the frequency by dividing the number of subplots with a specific plant type by the total number of subplots, which is 9 for the projects in this book.

Example

If six subplots have trees, the frequency of trees in the total test plot is calculated as follows:

$$\text{frequency} = \frac{6}{9} \times 100\% = 67\%$$

Appendix 5

Cross-Staff

Materials

pen
white copy paper
scissors

stapler
yardstick (meterstick)

Procedure

1. Trace or photocopy the crosspiece pattern (see Figure A5.1).
2. Cut along the solid lines.
3. Fold the paper away from you along the dashed lines.

CROSSPIECE

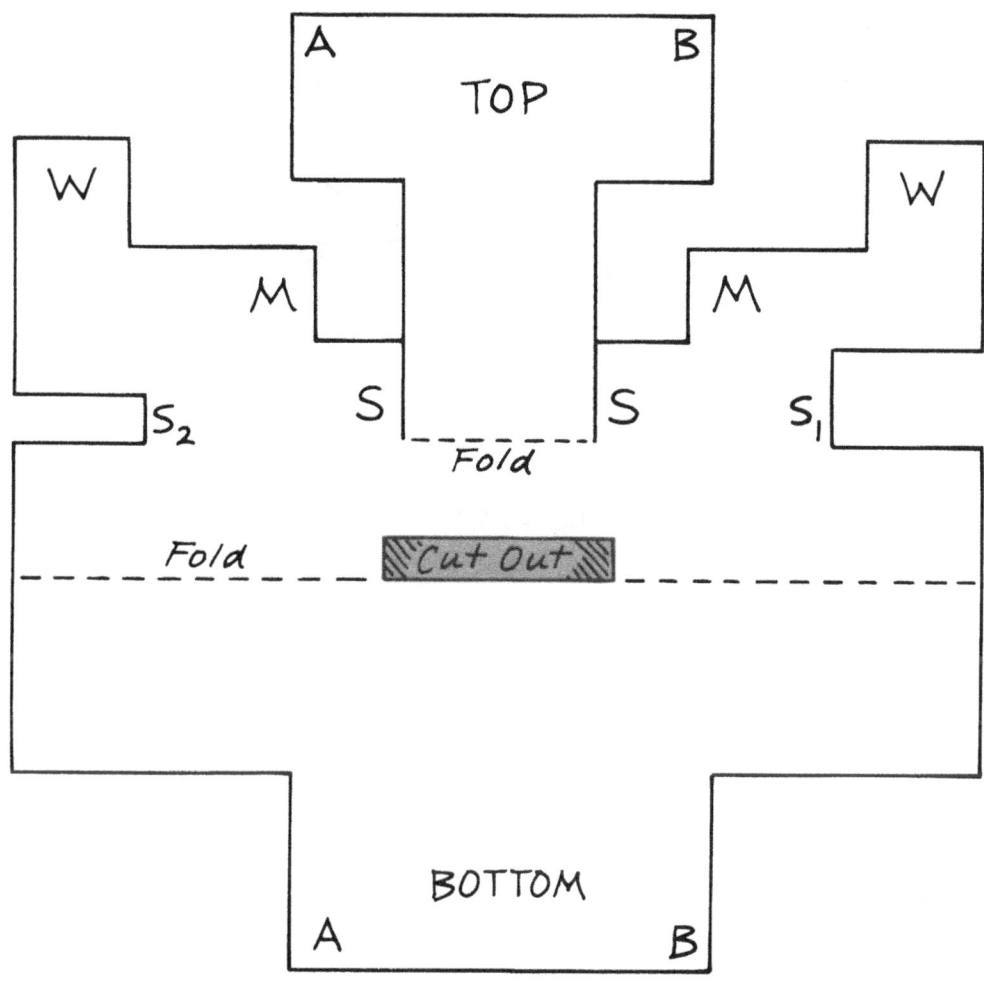

Figure A5.1

4. Position the T-shaped top so that its A and B sides meet the A and B sides of the bottom. Staple the A sides together first, then staple the B sides.

5. Slip the yardstick (meterstick) through the rectangular cutout section and between the top and bottom sections that are stapled together. The side labeled "W," "M," and "S" should face the zero end of the measuring stick. The paper is the crosspiece. It has a wide sight, "W," 4 inches (10 cm) across; a medium sight, "M," 2 inches (5 cm) across; and a small sight, "S," 1 inch (2.5 cm) across (see Figure A5.2). The notches in each side of the crosspiece are smaller sights and will be used in a later experiment.

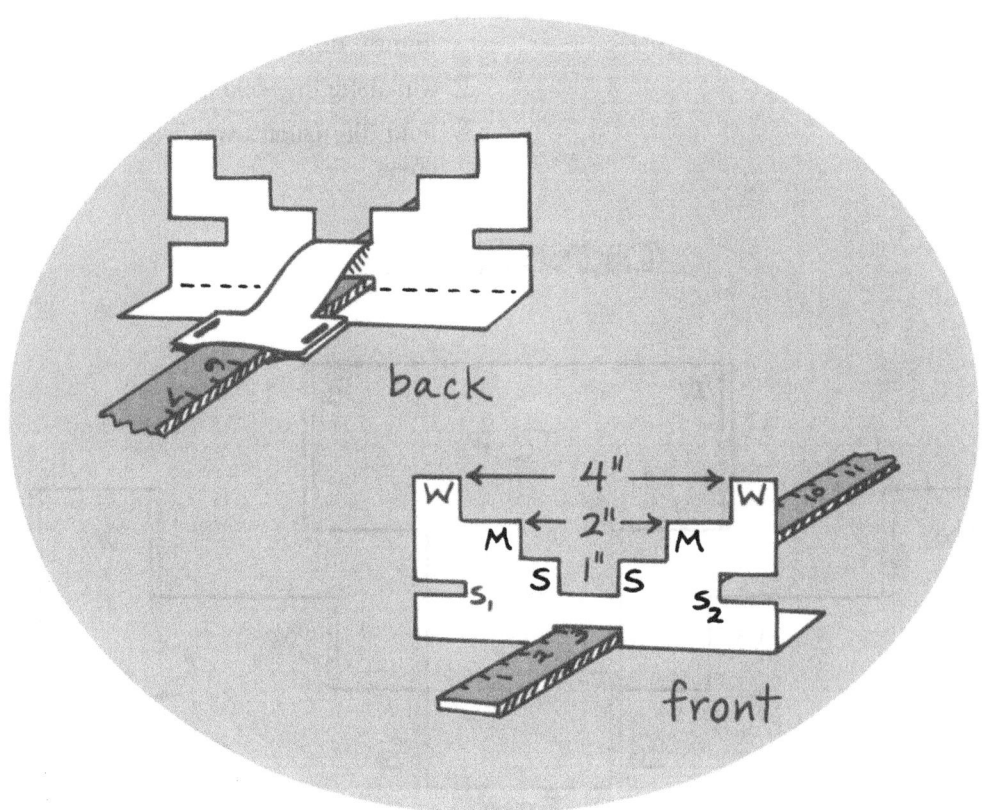

Figure A5.2

Appendix 6

Preparing Test Solutions

Red Cabbage Indicator

Materials

red cabbage
knife
cutting board
1-cup (250-ml) measuring cup
electric blender
4 cups (1000 ml) distilled water

large strainer
large bowl
tape
marking pen
refrigerator

Procedure

1. Cut enough red cabbage into small pieces to fill the cup.

2. Pour the cabbage pieces into the blender. Then add 4 cups (1000 ml) of distilled water.

3. Blend the water and cabbage.

4. Hold the strainer over the bowl and empty the contents of the blender into the strainer.

5. Pour the cabbage juice from the bowl into the jar and discard the solid pieces of cabbage left in the strainer.

6. Use the tape and marking pen to label the jar "Red Cabbage Indicator."

7. To prevent spoilage, store the indicator in a refrigerator until needed. It should be discarded after seven days.

Starch Solution (1%)

Materials

½ teaspoon (2.5 ml) cornstarch
1 cup (250 ml) distilled water

small saucepan
1-pint (500-ml) jar with lid

Procedure

1. Mix the starch with 1 tablespoon (15 ml) of the water to form a paste.

2. Boil the remaining water in the saucepan and slowly add the starch paste.

3. Cook for two minutes, stirring constantly. Allow to cool.

4. Pour into the jar and secure the lid.

5. To prevent spoilage, store the indicator in a refrigerator until needed. It should be discarded after seven days.

Figure A6.1

Brom Thymol Blue

Materials

1-quart (1-liter) jar with lid
1 quart (1 liter) of distilled
    water

0.1 gram of brom
    thymol blue

Procedure

1. Fill the jar with distilled water.

2. Add the brom thymol blue to the jar.

3. Secure the lid and shake the jar to mix. *Note:* If the solution appears green or yellow, add one drop of household ammonia (one drop at a time) until the indicator appears blue.

Calcium Hydroxide (Limewater)

Materials

2 1-quart (1-liter) glass jars with
    lids
distilled water
calcium oxide (lime used to
    make pickles)

masking tape
marking pen

Procedure

1. Fill one jar with distilled water.

2. Add 1 teaspoon (5 ml) of calcium oxide and stir.

3. Secure the lid and allow the solution to stand overnight.

4. Decant (pour off) the clear liquid into the second jar. Be careful not to pour any of the lime that has settled on the bottom of the jar.

5. Secure the lid and keep the jar closed.

6. Use masking tape and a marking pen to label the jar "Limewater."

Appendix **7**

Manometer

Materials

pencil
ruler
6-inch-by-30-inch (15-cm-by-75-cm) piece of poster board
transparent tape
scissors
9-inch (23-cm) round balloon
2- to 3-inch (5- to 7.5-cm) funnel

48-inch (120-cm) piece of aquarium tubing
cup
tap water
red food coloring
spoon
walnut-size piece of modeling clay

Procedure

1. Use the pencil and the ruler to mark lines across the poster board at 6 inches (15 cm) and 18 inches (45 cm) from one short side as shown in Figure A7.1.

2. Fold the poster board on these lines and tape the ends together to form a 3-sided stand as shown in Figure A7.2.

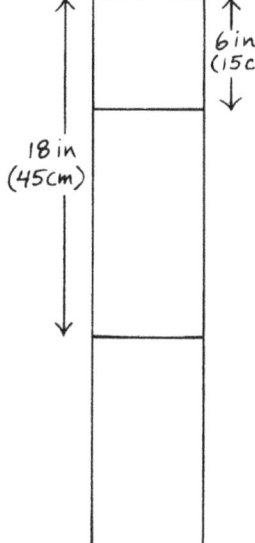

6 in (15cm)
18 in (45cm)

Figure A7.1

Figure A7.2

3. Tape the ruler vertically in the middle of one side of the stand.

4. Cut off the neck of the balloon. Discard the neck and stretch the bottom section of the balloon across the mouth of the funnel.

5. Fill the cup about one-fourth full with water.

6. Add about 20 drops of red food coloring to the cup of water. Stir.

7. Place one end of the aquarium tubing in the cup of colored water. Fill about 18 inches (45 cm) of the tube using your mouth to suck the water into the tube as you would draw liquid into a straw. Tap the tube to release any air bubbles in the water.

8. Tape about half of the tubing to the poster board so that it forms a U shape around the ruler as shown in Figure A7.3. Note the colored water is in the U shape of the tube.

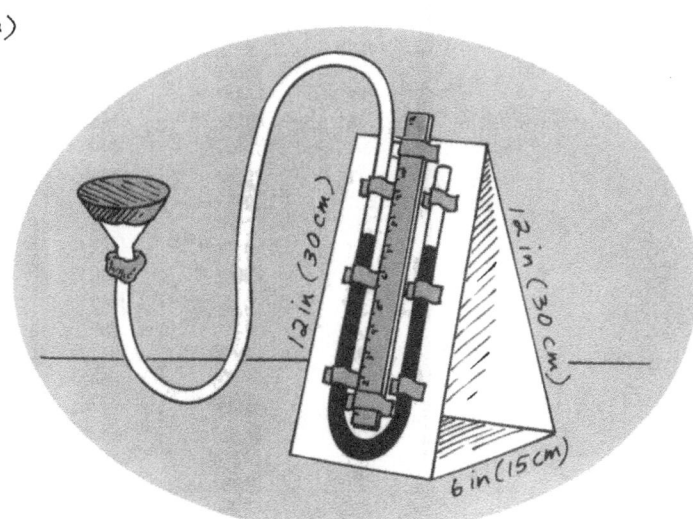

12 in (30cm)
12 in (30 cm)
6 in (15cm)

Figure A7.3

9. Insert the free end of the tubing into the end of the funnel and secure with modeling clay. You have made a manometer.

10. To use the manometer, place the balloon-covered funnel in an area where pressure is to be measured. A rise in the column of water on the open side of the U tube indicates an increase in pressure.

Appendix 8

Trigonometric Functions

Tangent Table					
Angle	cos	tan	Angle	cos	tan
0°	1.0000	.0000	45°	.7071	1.0000
1°	.9998	.0175	46°	.6947	1.0355
2°	.9994	.0349	47°	.6820	1.0724
3°	.9986	.0524	48°	.6691	1.1106
4°	.9976	.0699	49°	.6561	1.1504
5°	.9962	.0865	50°	.6428	1.1918
6°	.9945	.1051	51°	.6293	1.2349
7°	.9925	.1228	52°	.6157	1.2799
8°	.9903	.1405	53°	.6018	1.3270
9°	.9877	.1584	54°	.5878	1.3764
10°	.9848	.1763	55°	.5736	1.4281
11°	.9816	.1944	56°	.5592	1.4826
12°	.9781	.2126	57°	.5592	1.5399
13°	.9744	.2309	58°	.5299	1.6003
14°	.9703	.2493	59°	.5150	1.6643
15°	.9659	.2679	60°	.5000	1.7321
16°	.9613	.2867	61°	.4848	1.8040
17°	.9563	.3057	62°	.4695	1.8807
18°	.9511	.3249	63°	.4540	1.9626
19°	.9455	.3443	54°	.4384	2.0503
20°	.9397	.3640	65°	.4226	2.1445
21°	.9336	.3839	66°	.4067	2.2460
22°	.9272	.4040	67°	.3907	2.3559
23°	.9205	.4245	68°	.3746	2.4751
24°	.9135	.4452	69°	.3584	2.6051
25°	.9063	.4663	70°	.3420	2.7475
26°	.8988	.4877	71°	.3256	2.9042
27°	.8910	.5095	72°	.3090	3.0777
28°	.8829	.5317	73°	.2924	3.2709
29°	.8746	.5543	74°	.2756	3.4874
30°	.8660	.5774	75°	.2588	3.7321
31°	.8572	.6009	76°	.2419	4.0108
32°	.8480	.6249	77°	.2250	4.3315
33°	.8387	.6494	78°	.2079	4.7046
34°	.8290	.6745	79°	.1908	5.1446
35°	.8192	.7002	80°	.1736	5.6713
36°	.8090	.7265	81°	.1564	6.3138
37°	.7986	.7536	82°	.1392	7.1154
38°	.7880	.7813	83°	.1219	8.1443
39°	.7771	.8098	84°	.1045	9.5144
40°	.7660	.8391	85°	.0872	1.4301
41°	.7547	.8693	86°	.0698	14.3007
42°	.7431	.9004	87°	.0523	19.0811
43°	.7314	.9325	88°	.0349	28.6363
44°	.7193	.9657	89°	.0175	57.2900
45°	.7071	1.0000	90°	.0000	∞

Appendix 9

Science Project and Experiment Books

Astronomy

VanCleave, Janice. *A+ Astronomy*. New York: Wiley, 2001.

———. *Astronomy for Every Kid*. New York: Wiley, 1991.

Wood, Robert W. *Science for Kids: 39 Easy Astronomy Experiments*. Blue Ridge Summit, PA: TAB BOOKS, 1991.

Biology

Bonnet, Robert L., and G. Daniel Keen. *Botany: 49 Science Fair Projects*. Blue Ridge Summit, PA: TAB Books, 1989.

———. *Botany: 49 More Science Fair Projects*. Blue Ridge Summit, PA: TAB books, 1995.

Cain, Nancy Woodard. *Animal Behavior Science Projects*. New York: Wiley, 1995.

Dashefsky, H. Steven. *Zoology: 49 Science Fair Projects*. New York: Wiley, 1995.

Hershey, David R. *Plant Biology Science Projects*. New York: Wiley, 1995.

Kneidel, Sally. *Pet Bugs*. New York: Wiley, 1994.

Russo, Monica. *The Insect Almanac*. New York: Sterling, 1992.

VanCleave, Janice. *A+ Biology*. New York: Wiley, 1993.

———. *Animals*. New York: Wiley, 1993.

———. *Biology for Every Kid*. New York: Wiley, 1990.

———. *Foods and Nutrition for Every Kid*. New York: Wiley, 1999.

———. *The Human Body for Every Kid*. New York: Wiley, 1995.

———. *Insects*. New York: Wiley, 1998.

———. *Microscopes and Magnifying Lenses*. New York: Wiley, 1993.

———. *Plants*. New York: Wiley, 1997.

Wood, Robert W. *Science for Kids: 39 Easy Animal Biology Experiments*. Blue Ridge Summit, PA: TAB Books, 1991.

———. *Science for Kids: 39 Easy Plant Biology Experiments*. Blue Ridge Summit, PA: TAB Books, 1991.

Chemistry

Johnson, Mary. *Chemistry Experiments*. London: Usborn Publishing, 1981.

VanCleave, Janice. *Chemistry for Every Kid*. New York: Wiley, 1989.

———. *A+ Chemistry*. New York: Wiley, 1993.

———. *Molecules*. New York: Wiley, 1993.

Earth Science

Levine, Shar, and Allison Grafton. *Projects for a Healthy Planet*. New York: Wiley, 1992.

VanCleave, Janice. *A+ Earth Science*. New York: Wiley, 1999.

———. *Dinosaurs for Every Kid*. New York: Wiley, 1994.

———. *Earthquakes*. New York: Wiley, 1993.

———. *Earth Science for Every Kid*. New York: Wiley, 1991.

———. *Ecology for Every Kid*. New York: Wiley, 1995.

———. *Geography for Every Kid*. New York: Wiley, 1993.

———. *Oceanography for Every Kid*. New York: Wiley, 1995.

———. *Rocks and Minerals*. New York: Wiley, 1996.

———. *Solar Systems*. New York: Wiley, 2000.

———. *Volcanoes*. New York: Wiley, 1994.

———. *Weather*. New York: Wiley, 1995.

Wood, Robert W. *Science for Kids: 39 Easy Geology Experiments*. Blue Ridge Summit, PA: TAB Books, 1991.

———. *Science for Kids: 39 Easy Meteorology Experiments*. Blue Ridge Summit, PA: TAB Books, 1991.

Math

VanCleave, Janice. *Geometry for Every Kid*. New York: Wiley, 1994.

———. *Math for Every Kid*. New York: Wiley, 1991.

Physics

Amery, Heather, and Angela Littler. *The How Book of Batteries and Magnets*. London: Usborn Publishing, 1989.

Cobb, Vicki. *Science Experiments You Can Eat*. New York: HarperTrophy, 1994.

VanCleave, Janice. *A+ Physics*. New York: Wiley, 2002.

———. *Electricity*. New York: Wiley, 1994.

———. *Gravity*. New York: Wiley, 1993.

———. *Machines*. New York: Wiley, 1993.

———. *Magnets*. New York: Wiley, 1993.

———. *Physics for Every Kid*. New York: Wiley, 1991.

Wiese, Jim. *Roller Coaster Science*. New York: Wiley, 1994.

Wood, Robert W. *Sound Fundamentals*. New York: Learning Triangle Press, 1997.

General Project and Experiment Books

Amato, Carol J. *Super Science Fair Projects*. Chicago: Contemporary Books, 1994.

Bochinski, Julianne Blair. *The Complete Book of Science Fair Projects*, revised edition. New York: Wiley, 1996.

Bombaugh, Ruth. *Science Fair Success*. Hillside, NJ: Enslow, 1990.

Cobb, Vicki. *Science Experiments You Can Eat*. New York: HarperTrophy, 1994.

Frekko, Janet, and Phyllis Katz. *Great Science Fair Projects*. New York: Watts, 1992.

Levine, Shar, and Leslie Johnstone. *Everyday Science*. New York: Wiley, 1995.

———. *Silly Science*. New York: Wiley, 1995.

Markle, Sandra. *The Young Scientist's Guide to Successful Science Projects*. New York: Beech Tree Books, 1990.

Murphy, Pat, Ellen Klages, Linda Shore, and the Staff of the Exploratorium. *The Science Explorer*. New York: Henry Holt and Co., 1996.

Potter, Jean. *Science in Seconds for Kids*. New York: Wiley, 1995.

Smolinski, Jill. *50 Nifty Super Science Fair Projects*. Los Angeles: Lowell House Juvenile, 1995.

VanCleave, Janice. *Guide to the Best Science Fair Projects*. New York: Wiley, 1997.

———. *Guide to More of the Best Science Fair Projects*. New York: Wiley, 2000.

———. *200 Gooey, Slippery, Slimy, Weird, and Fun Experiments*. New York: Wiley, 1993.

———. *201 Awesome, Magical, Bizarre, and Incredible Experiments*. New York: Wiley, 1994.

———. *202 Oozing, Bubbling, Dripping, and Bouncing Experiments*. New York: Wiley, 1996.

Vacchione, Glen. *100 Amazing Make-It-Yourself Science Fair Projects*. New York: Sterling, 1994.

Wood, Robert W. *When? Experiments for the Young Scientist*. Blue Ridge Summit, PA: TAB Books, 1995.

Appendix **10**

Sources of Scientific Supplies

Catalog Suppliers

Carolina Biological Supply Company
2700 York Road
Burlington, NC 27215
(800) 334-5551

Cuisenaire
10 Bank Street
P.O. Box 5026
White Plains, NY 10606
(800) 237-3152

Delta Education, Inc.
P.O. Box 915
Hudson, NH 03051-0915
(800) 258-1302

Fisher Scientific
Educational Materials Division
485 South Frontage Road
Burr Ridge, IL 60521
(708) 655-4410
(800) 766-7000

Frey Scientific Division of Beckley Cardy
100 Paragon Parkway
Mansfield, OH 44903
(800) 225-3739

NASCO
901 Janesville Avenue
P.O. Box 901
Fort Atkinson, WI 53538
(800) 677-2960

Sargent-Welch
911 Commerce Court
Buffalo Grove, IL 60089
(800) 727-4368

Showboard
P.O. Box 10656
Tampa, FL 33679-0656
(800) 323-9189

Ward's Natural Science
5100 West Henrietta Road
Rochester, NY 14586
(800) 962-2660

Sources of Rocks and Minerals

The following stores carry rocks and minerals and are located in many areas. To find the stores near you, call the home offices listed below.

Mineral of the Month Club
1290 Ellis Avenue
Cambria, CA 93428
(800) 941-5594
cambriaman@thegrid.net
www.mineralofthemonthclub.com

Nature Company
750 Hearst Avenue
Berkeley, CA 94701
(800) 227-1114

Nature of Things
10700 West Ventura Drive
Franklin, WI 53132–2804
(800) 283-2921

The Discovery Store
15046 Beltway Drive
Dallas, TX 75244
(214) 490-8299

World of Science
900 Jefferson Road
Building 4
Rochester, NY 14623
(716) 475-0100

Project Journal

12 301

Printed and bound by CPI Group (UK) Ltd, Croydon, CR0 4YY

07/12/2025

14785991-0004